For Edexcel GCSE A

Tomorrow's Geography

FIFTH EDITION

Steph Warren

HODDER EDUCATION
AN HACHETTE UK COMPANY

In order to ensure that this resource offers high-quality support for the associated Pearson qualification, it has been through a review process by the awarding body. This process confirms that this resource fully covers the teaching and learning content of the specification or part of a specification at which it is aimed. It also confirms that it demonstrates an appropriate balance between the development of subject skills, knowledge and understanding, in addition to preparation for assessment.

Endorsement does not cover any guidance on assessment activities or processes (e.g. practice questions or advice on how to answer assessment questions), included in the resource nor does it prescribe any particular approach to the teaching or delivery of a related course.

While the publishers have made every attempt to ensure that advice on the qualification and its assessment is accurate, the official specification and associated assessment guidance materials are the only authoritative source of information and should always be referred to for definitive guidance.

Pearson examiners have not contributed to any sections in this resource relevant to examination papers for which they have responsibility.

Examiners will not use endorsed resources as a source of material for any assessment set by Pearson.

Endorsement of a resource does not mean that the resource is required to achieve this Pearson qualification, nor does it mean that it is the only suitable material available to support the qualification, and any resource lists produced by the awarding body shall include this and other appropriate resources.

Every effort has been made to trace all copyright holders, but if any have been inadvertently overlooked, the Publishers will be pleased to make the necessary arrangements at the first opportunity.

Although every effort has been made to ensure that website addresses are correct at time of going to press, Hodder Education cannot be held responsible for the content of any website mentioned in this book. It is sometimes possible to find a relocated web page by typing in the address of the home page for a website in the URL window of your browser.

Hachette UK's policy is to use papers that are natural, renewable and recyclable products and made from wood grown in well-managed forests and other controlled sources. The logging and manufacturing processes are expected to conform to the environmental regulations of the country of origin.

Orders: please contact Hachette UK Distribution, Hely Hutchinson Centre, Milton Road, Didcot, Oxfordshire, OX11 7HH. Telephone: +44 (0)1235 827827. Email education@hachette.co.uk Lines are open from 9 a.m. to 5 p.m., Monday to Friday. You can also order through our website: www.hoddereducation.co.uk

ISBN: 978 1471861253

© Steph Warren 2016

First published in 2016 by Hodder Education, An Hachette UK Company,
Carmelite House, 50 Victoria Embankment, London EC4Y 0DZ

www.hoddereducation.co.uk

This fifth edition published 2016

Impression number 11

Year   2025 2024

Cover photo © Oanh/Image Source/Corbis
Illustrations by Barking Dog Art and DC Graphic Design Ltd
Typeset by DC Graphic Design Ltd, Hextable, Kent
Printed by CPI Group (UK) Ltd, Croydon CR0 4YY

A catalogue record for this title is available from the British Library.

Contents

Acknowledgements

The publishers would like to thank the following for permission to reproduce copyright material:

Text acknowledgements: Inside back cover © Crown copyright 2016 Ordnance Survey. Licence number 100036470; **p.40** © Crown copyright 2016 Ordnance Survey. Licence number 100036470; **p.52** © Crown copyright 2016 Ordnance Survey. Licence number 100036470; **p.73** Wind roses © Crown copyright 2015, adapted from images supplied by Met Office: (http://www.metoffice.gov.uk/climate/uk/regional-climates); **p.135** eMarketer, February 2013, 'UK Ecommerce sales per person lead all markets worldwide, www.emarketer.com (http://www.emarketer.com/Article/UK-Ecommerce-Sales-per-Person-Lead-All-Markets-Worldwide/1009650); **p.152** Corruptions Perceptions Index 2014 © 2014 by Transparency International (https://www.transparency.org). Licensed under CC BY-ND 4.0. The Corruption Perceptions Index measures the perceived level of public sector corruption in countries and territories around the world; **p.159** Anup Shah, 'Foreign Aid for Development Assistance' (http://www.globalissues.org) (http://www.globalissues.org/article/35/foreign-aid-development-assistance) (last updated 28 September 2014), data taken from OECD (http://www.oecd.org/dac/stats/idsonline.htm) (accessed September 2014); **p. 169** Tanzania population, data taken from http://countrymeters.info/en/Tanzania; **p.173** Domestic Network Coverage, National ICR Broadband Backbone (http://www.nictbb.co.tz/map.php); **p.176** 'Commercial Fishing', BritishSeaFishing.co.uk (http://britishseafishing.co.uk/conservation/commercial-fishing/); **p.178** Jean-Marc Jancovici, 'Can't we put the carbon dioxide in a big hole and just forget it?' (http://www.manicore.com/anglais/documentation_a/greenhouse/sequestration_a.html) (last updated August 2013); **p.181** 'Toward a world of thirst?', Vital Water Graphics: An overview of the state of the world's fresh and marine waters, 2nd edition, 2008 (http://www.unep.org/dewa/vitalwater/); **p.205** 'Comparing global models of terrestrial net primary productivity (NPP): the importance of water availability', Global Change Biology, Vol. 5, Issue S1, pages 46–55, April 1999, G. Churkina et al.; **p.207** 'Freshwater use by sector at the beginning of the 2000s', Vital Water Graphics: An overview of the state of the world's fresh and marine waters, 2nd edition, 2008 (http://www.unep.org/dewa/vitalwater/article48.html); **p.219** Data taken from Environment Agency; **p.247** Long-term International Migration - Office for National Statistics.

Photo credits: p.1 © Stephen Dorey - Commercial/Alamy; **p.2** © Steph Warren; **p.3** all © Steph Warren; **p.4** all © Steph Warren; **p.8** © Lucy Phipps; **p.9** © Steph Warren; **p.10** © Steph Warren; **p.11** © Steph Warren; **p.16** © Steph Warren; **p.17** © Steph Warren; **p.18** all © Steph Warren; **p.19** © TOBY MELVILLE/Reuters/Corbis; **p.20** © Steph Warren; **p.21** all © Steph Warren; **p.22** all © Steph Warren; **p.24** © Skyscan/J Farmar; **p.26** © Steph Warren; **p.27** © Steph Warren; **p.30** all © Steph Warren; **p.32** © Steph Warren; **p.33** cr © Steph Warren; bl © Lucy Phipps; **p.35** © Steph Warren; **p.36** © Matt Cardy/Getty Images; **p.37** all © Steph Warren; **p.38** all © Steph Warren; **p.39** © GETMAPPING PLC/SCIENCE PHOTO LIBRARY; **p.41** © Steph Warren; **p.42** © Steph Warren; **p.45** all © Steph Warren; **p.48** © Skyscan/Air Images; **p.49** © Steph Warren; **p.50** © Janfrie1988/Public domain/https://commons.wikimedia.org/wiki/File:Abbey_Craig.JPG; **p.51** all © Steph Warren; **p.53** © Steph Warren; **p.54** © Steph Warren; **p.55** t © Steph Warren; **p.55** b © John McGovern/Alamy Stock Photo; **p.58** all © Steph Warren; **p.66** t ©azboomer55 - Fotolia; b © STC 11403, Houghton Library, Harvard University/https://commons.wikimedia.org/wiki/File:Houghton_STC_11403_-_Great_Frost,_1608.jpg; **p.68** © Steph Warren; **p.69** t © Steph Warren; b ©Dave Allen/ www.CartoonStock.com; **p.70** © NASA images by Jesse Allen and Robert Simmon, using Landsat 4, 5 and 7 data from the USGS Global Visualisation Viewer; **p.77** Original idea from Ros Hook; **p.79** l © STR/AFP/Getty Images; r © STR/AFP/Getty Images; bl © STR/AFP/Getty Images; br © Desmond Boylan/Reuters/Corbis; **p.80** tl © DOD Photo/Alamy Stock Photo; **p.80** tr © FEMA/Alamy Stock Photo; bl © Richard Levine/Alamy Stock Photo; br © Dan Callister/Alamy Stock Photo; **p.83** l © Larry Lilac/Alamy Stock Photo; r © Sabena Jane Blackbird/Alamy Stock Photo; **p.85** © Hanna Butler/IFRC; **p.86** b © imageBROKER/REX Shutterstock; **p.86** t © Hanna Butler/IFRC; **p.86** c © Hannah Butler/IFRC; **p.87** tl © David McNew/Getty Images; b © US Marines Photo/Alamy Stock Photo; **p.88** bl © David McNew/Getty Images; br © Lisa Werner/Alamy Stock Photo; **p.93** © Steph Warren; **p.96** all © Steph Warren; **p.98** © Paul Grover/Alamy Stock Photo; **p.99** © Steph Warren; **p.100** tl © ktsdesign - Fotolia; **p.100** cl © Agencja Fotograficzna Caro/Alamy Stock Photo; bl © Steph Warren; tr © Eduardo Rivero - Fotolia; cr © Steph Warren; br © Steph Warren; **p.103** l © BrazilPhotos.com/Alamy Stock Photo; r © Rhett A. Butler / mongabay.com; **p.104** © Sue Cunningham Photographic/Alamy Stock Photo; **p.105** © Steph Warren; **p.106** © Steph Warren; **p.107** © Steph Warren; **p.108** bl © John Foxx; t © WILDLIFE GmbH / Alamy Stock Photo; br © Imagestate Media (John Foxx)/Amazing Animals Vol 26; **p.111** © Steph Warren; **p.114** all © Steph Warren; **p.115** al © Steph Warren; **p.117** © www.webaviation.co.uk; **p.126** all © Steph Warren; **p.127** all © Steph Warren; **p.128** all © Steph Warren; **p.136** © Steph Warren; **p.137** all © Steph Warren; **p.140** © Public domain/https://commons.wikimedia.org/wiki/File:Rodovia_dos_Imigrantes_1.jpg; **p.141** tl © MAURICIO LIMA/AFP/Getty Images; br © Dornicke/https://commons.wikimedia.org/wiki/File:Bela_Vista_e_Liberdade_01.jpg/https://creativecommons.org/licenses/by-sa/4.0/deed.en; **p.142** © Diego Lezama Orezzoli/Corbis; **p.144** © David R. Frazier Photolibrary, Inc./Alamy Stock Photo; **p.145** t © Gili Yaari/NurPhoto; b © Paulo Whitaker/Reuters/Corbis; **p.146** © Noah Addis/Corbis; **p.148** t © Gavin Mather/Alamy Stock Photo; b © Paulo Fridman/Alamy Stock Photo; **p.149** cr © Anna Kari/In Pictures/Corbis; br © Sue Cunningham Photographic / Alamy Stock Photo; **p.150** © Gavin Mather/Alamy Stock Photo; **p.153** © Philip's; **p.154** © Philip's; **p.158** bl © Edward Parker/Alamy Stock Photo; **p.158** tr © Xinhua/Alamy Stock Photo; **p.158** tl © Norman Barrett/Alamy Stock Photo; **p.158** br © Susan Pease/Alamy Stock Photo; **p.161** © JTB MEDIA CREATION, Inc./Alamy Stock Photo; **p.177** © Scubazoo/Alamy Stock Photo; **p.179** © Philip's; **p.180** all © Philip's; **p.181** all © Philip's; **p.182** © Philip's; **p.183** © Philip's; **p.184** © Philip's; **p.186** © redbrickstock.com/Alamy Stock Photo; **p.187** © Steph Warren; **p.199** bl © Statkraft; br © Rehro/https://commons.wikimedia.org/wiki/File:Alta-damm.jpg/https://creativecommons.org/licenses/by-sa/3.0/de/deed.en; **p.200** b © DinodiaRF/Inmagine; **p.201** © EyesWideOpen/Getty Images; **p.204** © all Philip's; **p.208** t ©Photodisc/Getty Images; c © robertharding/Alamy Stock Photo; b © Steph Warren; **p.211** © Mirrorpix; **p.212** t © Lou DeMatteis/The Image Works/TopFoto; **p.212** b © Robert Nickelsberg/Getty Images; **p.215** © BEN STANSALL/AFP/Getty Images; **p.220** all © Steph Warren; **p.225** all © Steph Warren; **p.226** © Steph Warren; **p.229** all © Steph Warren; **p.244** all © Steph Warren; **p.246** © Steph Warren; **p.250** all © Steph Warren; **p.252** all © Steph Warren.

Introduction

This fifth edition of the popular Tomorrow's Geography series covers the subject content for the Edexcel GCSE Geography Specification A course which is first examined in 2018. The book covers the whole of the course, component by component, option by option as it is laid out in the specification.

Features of the textbook

The book contains all of the content for the specification but also has other helpful features. These include:

> **LEARNING OBJECTIVE**
>
> To study how Bristol is being changed by movements of people, employment and services.
>
> **Learning outcomes**
>
> ▶ To be able to describe the sequence of urbanisation, suburbanisation, counter-urbanisation and re-urbanisation processes and their distinctive characteristics for Bristol.
> ▶ To be able to explain the causes of national and international migration and their impact on different parts of Bristol.

Learning objectives and **learning outcomes** so that you know what you are expected to learn from each sub-topic in the book.

> **KEY TERMS**
>
> **Public buildings** – buildings owned by the council that serve the residents of the city, such as a library.

Key terms to remind you of the important words you will learn in this chapter.

> # Review
>
> By the end of this section you should be able to:
>
> ✓ understand how upland and lowland landscapes result from the interaction of physical processes

Review sections so that you can check the information that you have learned in this part of the book.

> **ACTIVITIES**
>
> 1 Which wards of Bristol have:
>   a) the most children
>   b) the least children?
> 2 Which age range in Bristol has the most people?
>   a) 0–15
>   b) 16–24
>   c) 25–49
>   d) 50–64
> 3 Calculate the percentage increase in recycling in Bristol between 2004 and 2012.
>
> Extension
> Assess the strategies which are used by Bristol Council to improve the quality of life for people who live in the city.

Activities which include calculations to help you prepare for the exam questions which require mathematical skills.

> **Fieldwork ideas** 💡
>
> **Suggested hypothesis:** the beach morphology on Swanage beach shows evidence of longshore drift occurring from south to north.
>
> **Primary fieldwork:**
> • measure the height of the sand on each side of the groynes

Fieldwork ideas on different aspects of the specification.

Extension activities to stretch the independent learners and provide extra information such as website addresses.

> **Practise your skills**
>
> Study Figure 1.7 on page 5. On a blank map of the UK:
>
> 1 Locate areas of each of the main rock types.
> 2 Name an area for each of the main rock types.

> **Examination-style questions**
>
> 1 Study Figure 1.7b on page 5. Identify the location of one area of schist landscape.   (1 mark)
> 2 State one example of igneous rock.   (1 mark)
> 3 State one characteristic of igneous rocks.   (1 mark)
> **Total: 3 marks**

Practise your skills questions to give you plenty of experience of the geography skills that are now part of the GCSE course. Some of these sections will be questions based on the OS maps provided in the textbook to ensure you do not forget essential OS map skills.

Examination-style questions which reflect the style of questions which will be found in the examinations to give you plenty of practice.

Course detail and book coverage

The table shows you how the book covers the GCSE course. The colours represent the colour coding of the chapters in the textbook.

Component	Exam section / Specification Topic	Options	Chapter	Assessment overview
Component 1 The Physical Environment Paper 1 code: 1GA0/01 Time allowed: 1 hour and 30 mins	Section A Topic 1: The Changing Landscapes of the UK	Two of the following three options in Topic 1: 1A: Coastal Landscapes 1B: River Landscapes 1C: Glaciated Upland Landscapes	Chapter 1 Chapter 2 Chapter 3 Chapter 4	The paper has three sections each worth 30 marks. 4 extra marks will be awarded for spelling, punctuation, grammar and use of specialist terminology, making a total for the paper of 94 marks. In Section A, students answer Question 1 and choose two others either on coasts, rivers or glaciated upland landscapes. Students answer all questions in Sections B and C. The sort of questions to expect are multiple choice, short and long open response, calculations and 8 mark extended writing questions.
	Section B Topic 2: Weather Hazards and Climate Change		Chapter 5	
	Section C Topic 3: Ecosystems, Biodiversity and Management		Chapter 6	
Component 2 The Human Environment Paper 2 code: 1GA0/02 Time allowed: 1 hour and 30 mins	Section A Topic 4: Changing Cities		Chapter 7 Chapter 8 Chapter 9	The paper has three sections each worth 30 marks. 4 extra marks will be awarded for spelling, punctuation, grammar and use of specialist terminology, making a total for the paper of 94 marks. Students answer all questions in Sections A and B. In Section C, students answer all of the questions on Resource management and choose one other, either on Energy or Water resource management. The sort of questions to expect are multiple choice, short and long open response, calculations and 8 mark extended writing questions.
	Section B Topic 5: Global Development		Chapter 10	
	Section C Topic 6: Resource Management	One of the following two options in Topic 6: 6A: Energy Resource Management 6B: Water Resource Management	Chapter 11 Chapter 12 Chapter 13	
Component 3 Geographical Investigations Paper 3 code: 1GA0/03 Time allowed: 1 hour and 30 mins	Sections A and B Topic 7: Geographical Investigations: Fieldwork	In Section A – students choose one from two optional questions on either rivers or coastal fieldwork. In Section B – students choose one from two optional questions on either central/inner urban areas or rural settlements.	Chapter 14	The paper has three sections. Section A and B are each worth 18 marks. Section C is worth 24 marks. 4 extra marks will be awarded for spelling, punctuation, grammar and use of specialist terminology, making a total for the paper of 64 marks. The sort of questions to expect are multiple-choice, short and long open response, calculations and 8 mark and 12 mark extended writing questions
	Section C Topic 8: Geographical Investigations: UK challenge	In Section C – students answer all questions.	Chapter 15	

Part 1 The Physical Environment

In the following chapters, you will study the content you need for Component 1: The Physical Environment.

This component is divided into three topics:

Topic 1 The Changing Landscapes of the UK

In this topic you will study **Chapter 1:** The Changing Landscapes of the UK and two of the following chapters:

Chapter 2: Coastal Landscapes and Processes

Chapter 3: River Landscapes and Processes

Chapter 4: Glaciated Upland Landscapes and Processes

Topic 2 Weather Hazards and Climate Change

In this topic you will study **Chapter 5,** an overview of the global circulation of the atmosphere and climate change over time, and look at tropical cyclones and drought.

Topic 3 Ecosystems, Biodiversity and Management

In this topic you will study **Chapter 6,** an overview of the distribution and characteristics of global and UK ecosystems and look in detail at tropical rainforests and deciduous woodlands.

The Changing Landscapes of the UK

KEY TERMS

Geology – the science that deals with the physical structure of the Earth, its history and how it changes.

Texture – the feel and appearance of a material.

Composition – what a material is made up of.

Fossils – the remnants of prehistoric organisms, such as a fish skeleton or a leaf imprint, which have become embedded in a rock.

There are geological variations within the UK

How did the UK's main rock types form and what are their characteristics?

The UK, although a small country, has a wide variety of landscapes. The geology of the UK has played a role in this variety. The main rock types found in the UK that will be discussed in this chapter are sedimentary (chalk and sandstone), igneous (basalt and granite) and metamorphic (schists and slates). These rocks display a number of distinctive characteristics.

Formation of sedimentary rocks

Sedimentary rocks are formed in layers. Many are formed from weathered or eroded rock debris that has been transported and deposited; the deposited rock grains build up in layers called sediments. The weight of the sediments cause the layers at the bottom to become compacted, forming sedimentary rocks such as sandstone. Other sedimentary rocks are formed in the same way; for example, dead sea creatures get compacted on the sea bed into chalk. This process can take millions of years.

> ### Characteristics of sedimentary rocks
> Sedimentary rocks:
> - are classified by texture and composition
> - usually have layers
> - often contain fossils
> - are composed of rounded grains pushed together
> - have a great variety in colour
> - are made of particles that may be the same size or vary.

⬆ **Figure 1.1** Sandstone at Baggy Point, Devon.

↑ **Figure 1.2** Chalk cliffs in Kent.

Formation of igneous rocks

Igneous rocks are formed from molten rock called **magma** that is found inside the Earth. When magma cools it forms igneous rocks. If magma cools underground, it cools slowly, forming rocks that contain large crystals such as granite. If magma erupts from a volcano, it cools quickly, forming rocks that contain small crystals, as basalt.

Characteristics of igneous rocks
Igneous rocks:

- are formed from molten rock (magma)
- are made from randomly arranged crystals
- are very resistant rocks
- do not contain fossils
- may be intrusive, forming inside the Earth, such as granite
- may be extrusive, forming on the Earth's surface, such as basalt.

↑ **Figure 1.3** Granite scenery at Haytor on Dartmoor.

↑ **Figure 1.4** Drumadoon basalt columns on the Isle of Arran, Scotland.

Formation of metamorphic rocks

These rocks form when igneous or sedimentary rocks are put under great pressure or are close to a source of heat. The rocks are not melted but are heated. Under these two conditions the minerals within the rock change chemically to form a new type of **metamorphic** rock.

Characteristics of metamorphic rocks

Metamorphic rocks:

- are formed from other rocks, either sedimentary or igneous
- are formed under great heat or pressure
- have crystals that can be arranged in layers, for example slate, which is formed from shale
- can contain fossils, although the fossils are usually squeezed out of shape, for example marble.

⬆ **Figure 1.5** Mica schists in south Devon.

⬆ **Figure 1.6** Slate extraction in Snowdonia.

Practise your skills

Study Figure 1.7 on page 5. On a blank map of the UK:

1 Locate areas of each of the main rock types.
2 Name an area for each of the main rock types.

The role of geology and past tectonic processes in the development of upland and lowland landscapes

The different types of rocks have varying resistance to physical processes. Igneous and metamorphic rocks tend to be more resistant and therefore form upland, or highland, areas. The igneous and metamorphic rocks in the UK were formed when it had **tectonic** activity. Volcanic cones can still be seen in the UK landscape, for example Abbey Craig near Stirling is built on a volcanic plug (see page 50). The island of Ailsa Craig is also a volcanic plug.

The lowland landscapes are formed from sedimentary rocks. These landscapes are not necessarily flat – they can contain rolling hills, such as the North Downs – but they are much lower landscapes as the rock types are less resistant to physical processes.

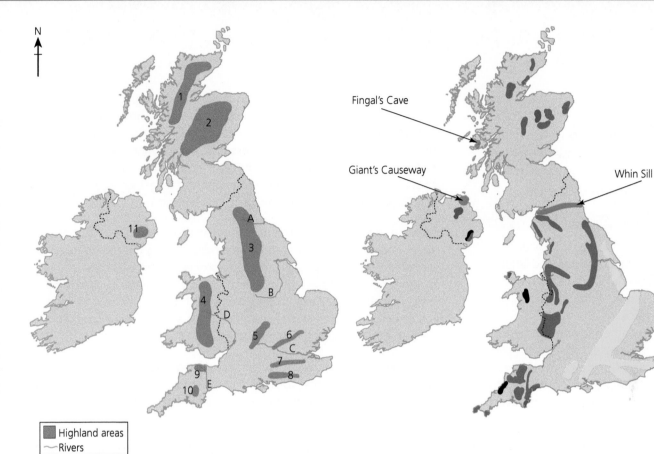

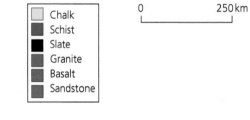

Location on map	Name of river
A	River Tees
B	River Trent
C	River Thames
D	River Severn
E	River Exe

Location on map	Name of upland
1	Northwest Highlands
2	Grampian Mountains
3	Pennines
4	Cambrian Mountains
5	Cotswolds
6	Chilterns
7	North Downs
8	South Downs
9	Exmoor
10	Dartmoor
11	Mourne Mountains

Highland areas
Rivers

Chalk
Schist
Slate
Granite
Basalt
Sandstone

0   250 km

⬆ **Figure 1.7a** UK upland areas.

⬆ **Figure 1.7b** Simplified distribution of rock types in the UK.

ACTIVITIES

1 Identify the location of one area of granite landscape in the UK.
2 Is basalt found in the southeast of England?
3 Name two areas of chalk hills in the UK.
4 Name a volcanic cone that can still be seen in the UK.

Extension

Visit the Geological Society's website (www.geolsoc. org.uk) to learn about the distribution of the UK's main rock types.

Review

By the end of this section you should be able to:

✓ describe the characteristics of the UK's main rock types
✓ locate the main rock types on a map of the UK
✓ understand the role of geology and past tectonic processes in the development of upland and lowland landscapes.

A number of physical and human processes work together to create distinct UK landscapes

◻ LEARNING OBJECTIVE

To study the physical and human processes that have created the distinctive landscapes of the UK.

Learning outcomes

▶ To understand how distinctive upland and lowland landscapes result from the interaction of physical processes (glacial erosion and deposition, weathering and climatological processes, post-glacial river and slope processes).
▶ To recognise how distinctive landscapes result from human activity (agriculture, forestry, settlement) over time.

The upland landscapes of the UK could not be described as mountainous but they are very different from the lowland areas. The upland areas were formed by resistant rocks millions of years ago and their landscape has been defined by the physical processes at work during the last ice age. The UK has been covered by ice during ice ages on a number of occasions. The extent of the coverage during the last ice age, about 20,000 years ago, is shown on Figure 1.8 on page 7.

Since the last ice age, weathering, climate and other physical agents have had an impact on the landscape. The landscape has been shaped by the work of rivers (see page 26) and slope processes, such as mass movement (see page 27). The climate of the UK has also had an impact on the landscape. For example, heavy rainfall will cause rivers to have greater erosive power. Mechanical, biological and chemical weathering (see page 8) also have a continual impact on the landscape.

The lowland areas of the UK were shaped by glacial outwash. As the glaciers melted the water formed distinctive lowland landscapes. These landscapes have continually been shaped by rivers, weathering and slope processes, outlined in Chapters 2, 3 and 4.

These landscapes have also changed over time due to human activity. The building of settlements is perhaps the most distinctive change: houses, industries and roads connecting settlements have changed the landscape forever, with natural landscapes becoming human ones.

The agricultural landscape also continues to change. Originally the land was farmed with hedges and walls as field boundaries; as farming practices have changed over time, however, the hedges in some parts of the country have been removed and extensive areas of land have been created to allow for the large machinery that is now used. The landscape of the countryside, particularly in lowland areas, is continually evolving.

The landscape of the UK has also been influenced by forestry. Originally much of the UK was covered by **deciduous** woodland. Over hundreds of years the woodland has been felled, which has changed the landscape in a number of ways. Land that was once covered in trees is now open moorland, settlement and farmland. There has also been a change in the *type* of woodland in the UK, because much of the woodland which has been grown to replace the felled woodland is **coniferous**, not deciduous. In Scotland, the amount of woodland had decreased to four per cent of the landscape by 1900. The total is now back up to almost twenty per cent.

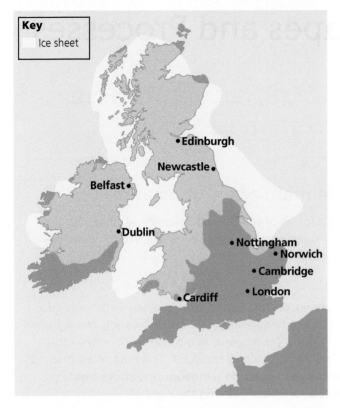

Key
Ice sheet

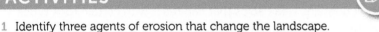

← **Figure 1.8** The extent of the ice sheet during the most recent glaciation – Devensian.

ACTIVITIES

1 Identify three agents of erosion that change the landscape.
2 Describe how rivers can change the landscape.
3 State three human activities that change the landscape.
4 Describe how farming has changed the landscape of the UK over the past 200 years.

Extension

Use the internet to research information about the last ice age. Try to answer the following questions.

1 What is meant by an ice age?
2 When was the last ice age?
3 Which areas of the UK were covered with ice?
4 What was the impact of the ice on the UK landscape of today?

Review

By the end of this section you should be able to:

✓ understand how upland and lowland landscapes result from the interaction of physical processes
✓ recognise that distinctive landscapes can result over time from human activity.

Examination-style questions

1 Study Figure 1.7b on page 5. Identify the location of one area of schist landscape.   (1 mark)
2 State one example of igneous rock.   (1 mark)
3 State one characteristic of igneous rocks.   (1 mark)
4 Forestry is a human activity that affects the landscape. Name one other human activity that affects the landscape.   (1 mark)
5 Explain how that activity affects the landscape.   (2 marks)

Total: 6 marks

2 Coastal Landscapes and Processes

LEARNING OBJECTIVE

To study the physical processes that interact to shape the coast.

Learning outcomes

▶ To understand the impact of weathering, mass movement and erosion on the coast.
▶ To understand the ways that the sea transports and deposits material along the coast.
▶ To be able to explain the process of longshore drift.
▶ To recognise the influence of geological structure, joints and faults, and rock type on landforms.
▶ To be able to identify concordant and discordant coastlines and recognise their influence on landforms.
▶ To know the characteristics of destructive and constructive waves and their influence on landforms.
▶ To know how the UK's weather and climate affect rates of coastal erosion and impact on coastal landscapes.

A variety of physical processes interact to shape coastal landscapes

What are the main types of weathering?

There are three main forms of **weathering**: mechanical, chemical and biological. Not all of them occur on every coastline but combinations of them are usually evident.

Mechanical weathering

Freeze–thaw weathering, or frost action, occurs when water gets into cracks in rocks. When the temperature falls below freezing, the water will expand as it turns into ice. This expansion puts pressure on the rock around it and fragments of rock may break off. This type of weathering is common in highland areas where the temperature is above freezing during the day and below freezing during the night.

Chemical weathering

Rainwater contains weak acids that can react with certain rock types. Carbonates in limestone are dissolved by these weak acids and this causes the rock to break up or disintegrate. This can be seen on limestone statues and limestone pavements.

Biological weathering

This is the action of plants and animals on the land. Seeds that fall into cracks in rocks will start to grow when moisture is present. The roots the young plant puts out force their way in and, in time, can break up rocks (see Figure 2.1). Burrowing animals, such as rabbits, can also be responsible for the further break-up of rocks. This is due to the way that they tunnel through the upper layers of the soil.

⬆ **Figure 2.1** Biological weathering.

What is mass movement?

Mass movement is when material moves down a slope due to the pull of gravity. There are many types of mass movement but, for the purposes of this chapter, only slumping and sliding will be discussed. Slumping, also known as rotational slipping, involves a large area of land moving down a slope. It is very common on clay cliffs: during dry weather the clay contracts and cracks; when it rains, the water runs into the cracks and is absorbed until the rock becomes saturated (see Figure 2.2). This weakens the rock and, due to the pull of gravity, it slips down the slope on its slip plane. Due to the nature of the slip, it leaves behind a curved surface.

⬆ **Figure 2.2** Slumping east of Bowleaze Cove, Dorset.

How are coasts eroded?

The coast is a narrow strip between land and sea. It is under continual attack from waves at the base of the cliff and other processes on the cliff face, such as weathering and mass movement. (You should always refer to these processes when answering a question on landform formation.) The theory box should be referred to throughout this chapter to understand the processes of **erosion**.

How does the sea transport materials?

Waves can transport materials in a number of ways, including: traction, saltation, suspension, solution and longshore drift.

- **Traction** – large sediment such as pebbles roll along the sea bed.
- **Saltation** – small pieces of shingle or large grains of sand are bounced along the sea bed.
- **Suspension** – small particles such as sand and clays are carried in the water; this can make the water look cloudy, especially during storms or when the sea has lots of energy.
- **Solution** – some minerals dissolve in sea water and are carried in solution. This is particularly evident near to limestone or chalk cliffs where the sea can appear to be a milky colour due to the amount of sediment being carried in solution.

Processes of coastal erosion

Abrasion

Sand and pebbles carried within waves are thrown against the cliff face with considerable force; these particles break off more rocks which, in turn, are thrown against the cliff by the breaking waves.

Hydraulic action

This is the pressure of the water being thrown against the cliffs by the wave. It also includes the compression of air in cracks: as the water gets into cracks in the rock face, it compresses the air in the cracks; this puts even more pressure on the cracks and pieces of rock may break off.

Solution

This is a chemical reaction between certain rock types and the salt and other acids in sea water. This is particularly evident on limestone and chalk cliffs where the water is a milky blue at the bottom of the cliffs due to the dissolved lime.

Attrition

This process involves the wearing away of the rocks that are in the sea. As the boulders in the sea continually roll around, they chip away at each other until smooth pebbles or sand are formed.

KEY TERMS

Swash – the forward movement of a wave.

Backwash – the movement of a wave back down the beach.

The process of longshore drift

Longshore drift is the movement of sand and pebbles down a coastline. The direction of the waves hitting the coastline is determined by the prevailing wind; if the wind is blowing at an angle to the beach, the waves (swash) will approach the beach at this angle, transporting the sand and pebbles with them. As the returning wave (backwash) is being pulled by gravity, it will take the shortest route back down the beach; it always goes back down the beach in a straight line at 90° to the coast. In this way, material is moved along the beach until it meets an obstruction, see Figure 2.3.

Fieldwork ideas

Suggested hypothesis: the beach morphology on Swanage beach shows evidence of longshore drift occurring from south to north.

Primary fieldwork:

- measure the height of the sand on each side of the groynes
- do beach profiles on different parts of the beach to ascertain different slopes

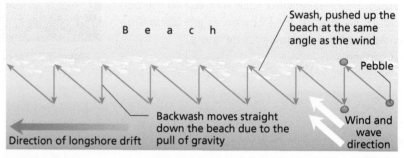

B e a c h

Swash, pushed up the beach at the same angle as the wind

Pebble

Backwash moves straight down the beach due to the pull of gravity

Wind and wave direction

Direction of longshore drift

⬆ **Figure 2.3** The process of longshore drift.

What is deposition?

Deposition is the laying down of materials, such as sand and pebbles, which are being transported by the sea. The sea will deposit materials when it slows down and loses energy, such as waves on a beach.

How does geological structure have an impact on landforms?

A rock's structure can affect the rate of erosion and the landforms that are produced. Rocks that are well jointed or have many faults, such as limestone, will erode more quickly as the waves exploit these lines of weakness. Rocks that have few joints will be harder for the sea to erode. Rocks such as chalk have lines of weakness known as bedding planes that allow the sea to erode them more easily (see Figure 2.4).

Bedding plane

Stack

Lines of weakness in the rock

Arch

⬆ **Figure 2.4** Chalk cliffs with bedding planes, Old Harry Rocks at Handfast Point, near Swanage, Dorset.

The type of rock on a coastline also affects the rate of erosion. Cliffs made from resistant rock, such as granite, will erode more slowly than cliffs made from less resistant rock, such as clay.

What are concordant and discordant coastlines?

Concordant coastlines have rocks that lie parallel to the coastline. **Discordant** coastlines have bands of rocks that lie at right angles to the coast. These geological structures influence the formation of different coastal landforms.

Concordant coastlines have alternate layers of hard (more resistant) and soft (less resistant) rock. The hard rock will act as a barrier to the erosive power of the sea. If the sea erodes through the hard rock it will then quickly erode the softer rock behind, as in the case of Lulworth Cove in Dorset.

Discordant coastlines have rocks that are at right angles to the sea. If there are alternate layers of hard and soft rock, the soft rock will erode more quickly forming bays with the hard rock forming headlands (see Figure 2.9 on page 13).

⬆ **Figure 2.5** Lulworth Cove, Dorset.

What types of waves are there?

Destructive waves

Destructive waves are the most important agent in coastal erosion and in taking sediment away from coastlines. Landforms produced by destructive waves include headlands, bays, caves, arches, cliffs, **stacks** and **wave-cut platforms**.

Destructive waves have a number of characteristics:

- The backwash is much stronger than the swash and is therefore able to carry sand and pebbles away from the shore.
- They break frequently; there are between ten and fifteen every minute.
- They are high in proportion to their length.
- They are generally found on steep beaches.

Constructive waves

Constructive waves are responsible for deposition in coastal areas and landforms such as beaches, bars and spits. They have a number of characteristics:

- The swash is more powerful than the backwash and therefore deposits sediment on beaches.
- They break infrequently, at a rate of ten or fewer per minute.
- They are long in relation to their height.
- They are usually found on gently sloping beaches.

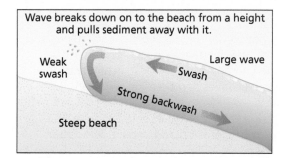

⬆ **Figure 2.6** A destructive wave.

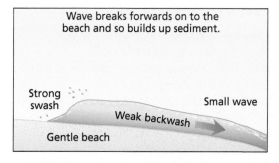

⬆ **Figure 2.7** A constructive wave.

How do the UK's weather and climate affect rates of coastal erosion?

KEY TERMS

Seasonality – a pattern of change in the UK's weather between spring, summer autumn and winter.

Fetch – the distance over which the wind blows over open water.

The seasonality of the UK's weather and climate affects the rate of coastal erosion. In the winter the differences between day and night-time temperatures can cause freeze–thaw weathering on cliff faces. Storms also have an impact on the landforms of the coastline as storm waves are powerful agents of erosion. The human coastal landscape, such as sea defences, is in need of constant repair due to the increasing regularity of storms.

The prevailing wind in the UK is from the southwest. The coastlines of Cornwall and Devon experience winds that may have blown for several thousand kilometres across the Atlantic Ocean (see Figure 2.8). These winds have a long fetch: the longer the fetch, the stronger the wind and the more powerful the wave, and the faster the rate of erosion.

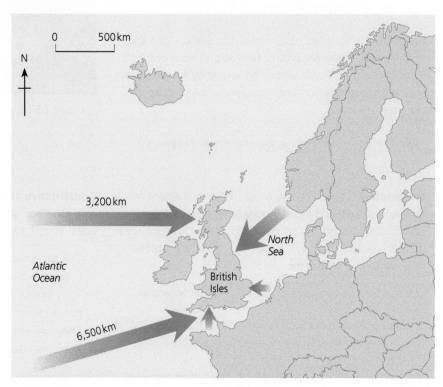

⬆ **Figure 2.8** The fetch of waves around the British coastline.

Review

By the end of this section you should be able to:

✓ describe the impact of weathering, mass movement and erosion on the coast
✓ describe the ways that the sea transports and deposits material along the coast
✓ explain the process of longshore drift
✓ know the influence of geological structure, joints and faults and rock type on landforms
✓ identify concordant and discordant coastlines and recognise their influence on landforms
✓ know the characteristics of destructive and constructive waves and their influence on landforms
✓ know how the UK's weather and climate affect rates of coastal erosion and impact on coastal landscapes.

Coastal erosion and deposition create distinctive landforms within the coastal landscape

LEARNING OBJECTIVE

To study the distinctive landforms created by coastal erosion and deposition.

Learning outcomes

▶ To be able to describe and explain the formation of headlands and bays, cliffs and wave-cut platforms, caves, arches, stacks and stumps.
▶ To be able to describe and explain the formation of beaches, spits and bars.

What landforms are created by coastal erosion?

Distinctive and dynamic landforms are formed by destructive waves. These include headlands and bays; cliffs and wave-cut platforms; caves, arches, stacks and stumps.

Headlands and bays

On coastlines where rocks of varying resistance lie at right angles to the sea, bays – indentations in the land – show where the softer rock is. Headlands are the more resistant rock and protrude into the sea. As the bays are made from a less-resistant rock type, the erosion rates from processes such as abrasion are greatest at first. In time, as the sea cuts the bays back, the waves reaching the coast are less powerful because they have to travel over a longer expanse of beach. At this point the headlands, which are further out to sea, start to experience the more powerful waves and are eroded at a faster rate than before, see Figure 2.9.

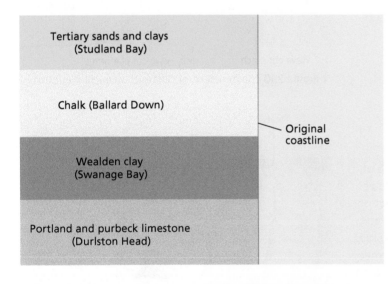

KEY TERMS

Dynamic landform – a landform that is changing.

Therefore, the features of the area, such as:
● the differing resistance of the rocks forming a discordant coastline
● the physical processes in the area such as erosion by the sea on the softer clay forming bays
● and eventually the use of hydraulic action on the headlands such as Ballard Down have formed the characteristics of the landscape seen today. Other processes such as physical weathering attack the cliffs causing them to recede further. At the same time deposition is occurring in the bays to form beaches.

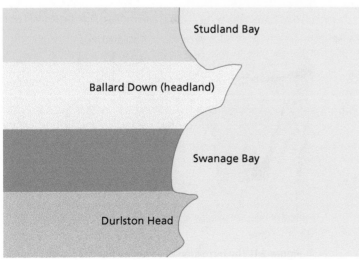

◀ **Figure 2.9** The formation of headlands and bays.

Cliffs and wave-cut platforms

Headlands are usually formed from cliffs. When the sea moves against the base of the cliff, using abrasion and hydraulic action (and solution if the rock type is limestone or chalk – see page 9), it undercuts the cliff and forms a wave-cut notch. An overhang will form above this notch which, in time, will fall into the sea as a result of the pressure of its own weight and the pull of gravity.

The sea will then continue to attack the cliff and form another notch. In this way, the cliff will retreat, becoming higher and steeper. The remains of the cliff rock, now below the sea at high tide, form a rocky, wave-cut platform. As a result of erosion and weathering, some boulders will have fallen from the cliff on to the platform. As the width of the platform increases, so the power of the waves decreases, as they have further to travel to reach the cliff, see Figure 2.10.

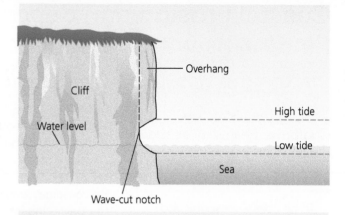

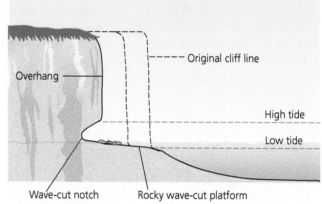

⬆ **Figure 2.10** The formation of cliffs and wave-cut platforms.

Caves, arches, stacks and stumps

These are formed in rocks that have a fault or line of weakness. The action of the sea will exploit the fault through erosional processes, such as hydraulic action. In time the fault will widen to form a cave. If the fault is in a headland, caves are likely to form on both sides. When the backs of the caves meet, an arch is formed. The sea will continue to erode the bottom of the arch using abrasion. As the sea undercuts the bottom of the arch, a wave-cut notch will form, which will collapse in time as it is pulled down by the pressure of its own weight and gravity. This leaves behind a column of rock not attached to the cliff, known as a stack. Continued erosion will lead to the formation of a stump that is visible only at low tide, see Figure 2.11.

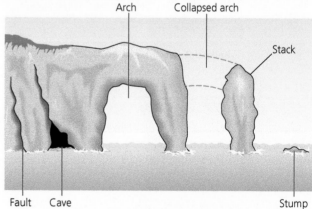

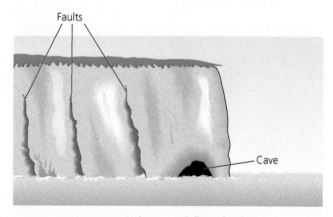

⬆ **Figure 2.11** The formation of caves, arches, stacks and stumps.

What landforms are created by coastal deposition?

Constructive waves build rather than destroy the coastal environment. They deposit sand and pebbles that form beaches, spits and bars.

Beaches

Beaches are perhaps the most easily recognised and named coastal feature around the British coast. A beach is an area of land between the low tide and storm tide marks and is made up of sand, pebbles and, in some places, mud and silt. They are formed by constructive waves, often in bays where the waves have less energy due to the gently sloping sea bed, and as a result deposit material. They can also be found along straight stretches of coastline where longshore drift occurs. Seaside resorts often build **groynes** to keep beaches in place and to reduce the effects of longshore drift.

Spits

A spit is a long, narrow stretch of pebbles and sand that is attached to the land at one end, with the other end tapering into the sea. It forms when longshore drift (see explanation on page 10) occurs on a coastline. When the coastline ends, the sea deposits the material it is transporting because the change in depth affects its ability to transport the material further.

If there is a river estuary, the meeting of the waves and the river causes a change in speed, which results in both the waves and the river dropping their sediment. In time, the material builds up to form a ridge of shingle and sand known as a spit. On the land side, silt and alluvium are deposited and salt marshes form. The wind and sea currents may curve the end of the spit around. Spits are very dynamic, which means that their shape and form continually change. If spits are present on a coastline, it should be possible to determine the direction of longshore drift (see Figure 2.12).

The spit in Figure 2.12 has formed because the characteristics of the area were ideal for this landform to develop. Longshore drift is occurring along the beach from west to east with the predominant wind direction being from the south-west. Longshore drift is the movement of the swash going up the beach at an angle pushed by the wind and the backwash bringing the sediment straight back down the beach due to the pull of gravity. The coast changes to a north easterly direction. The sea continues to move in a south-westerly direction driven by the wind. It will continue to move the

sand, depositing it to build a spit off the end of the coastline. Material will be deposited because the water will become deeper as the waves are further from the coastline. There will be a change in speed of the water (it will move more slowly) causing it to drop what it is carrying. The spit has a number of recurved ends which have formed when the wind or tide have changed direction.

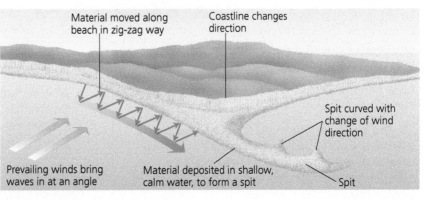

▲ **Figure 2.12** The formation of a spit.

Bars

If a spit develops in a bay, it may build across it, linking two headlands to form a bar. This is only possible if there is a gently sloping beach and no powerful river entering the sea. In this way, bars can straighten coastlines. An example is at Slapton in Devon, see Figure 2.13, which has the characteristic lagoon formed behind the bar where any water from streams is trapped and slowly seeps through the pebbles and stones of the bar into the sea. The bar was formed by the process of longshore drift which occurs on this coastline.

A lagoon formed by the bar and the small streams flowing into this area

A bar joined to the coastline at both ends

➡ **Figure 2.13**
A bar in Devon.

Review

By the end of this section you should be able to:

✓ describe and explain the formation of headlands and bays; cliffs and wave-cut platforms; caves, arches, stacks and stumps

✓ describe and explain the formation of beaches, spits and bars.

Practise your skills

Draw annotated diagrams of coastal erosion and deposition features.

Choose a coastal area close to where you live (or use the Isle of Purbeck, which is the distinctive coastal landscape featured in this book). Look at weather statistics for different years compared to erosion rates on the coastline.

ACTIVITIES

1 Match the term with its correct definition.

Term	Definition
Abrasion	The wearing away of rocks that are in the sea.
Solution	The wearing away of cliffs by the rocks in the sea.
Attrition	A chemical reaction between certain rock types and sea water.

2 List three differences between constructive and destructive waves.

3 What is meant by the term 'fetch'?

4 Explain the process of longshore drift.

5 Examine how physical processes work together to form the cliff and wave-cut platform in Figure 2.10.

Extension

Research the formation of the bar at Slapton in Devon.

Human activities can lead to changes in coastal landscapes that affect people and the environment

How have human activities such as urbanisation, agriculture and industry affected coastal landscapes?

The building of towns and cities on the coastline of the UK affects the landscape in many ways, from the visual impact of the settlements to the impact on wildlife and on the natural processes at work.

Urban areas developed on the coast for a number of reasons, such as fishing or trading purposes, which led to the construction of harbours and larger ports. As settlements grew, the coastal landscape changed. Original wetland areas were drained to ensure that the settlements were not flooded, meaning that wading birds and other animals lost their habitat. Harbours were built to give shelter to fishing boats, which involved the building of jetties into the cliffs. Large ports developed to import materials; these were usually on large estuaries, which had a major visual impact on the area as well as causing environmental changes. During the early twentieth century, tourist developments along the coast led to further change in the coastal landscape. All of these developments require defending against the sea. This in turn impacts on the natural processes that occur in coastal areas; for example, the building of groynes interferes with the process of longshore drift.

Parts of the coastal landscape are still mainly farming areas. Such areas are less developed, meaning that the coastal landscape is largely unchanged and natural processes are not interfered with, because the land is not seen as valuable enough to defend. However, many lowland coastal landscapes that were originally wetland areas have been drained and used for farming. This has had an environmental impact on the coastal landscape. For example, Cuckmere Haven in Sussex was drained for farming, which changed the area from a wetland to sheep farming. However, the coastal defences are no longer being repaired and the area will return to salt marsh as the sea gradually reclaims it. Another example is Porlock Bay in Somerset, where a shingle ridge was built to defend low-lying farmland from the sea. As the shingle is no longer being replenished, the bank has been breached by the sea, forming a salt marsh (see Figure 2.14).

Shingle ridge no longer defended from the sea

Farmland, which is flooded at high tide, becoming a salt marsh

⬆ **Figure 2.14** Porlock Bay, Somerset.

What are the effects of coastal recession on people and the environment?

England has 2,800 miles of coast, of which 1100 are classed as being at risk from erosion. Coastal recession affects both people and the environment. If the recession is occurring where there is a settlement then its effects on people will be greatest. If people lose their homes it will have a major impact both on their daily lives and their finances because they will have lost the money they had invested in their home. An example of this is Seaton in Devon, which will not be defended after 2025 because it is a small town. This means that, over time, the houses, pub and campsite will be lost to the sea due to coastal recession. Other impacts on people could be the effects on farmers who lose their land to the sea. This could impact on the viability of their farm as it loses fields where crops were once grown. Coastal recession also affects transport networks, making it difficult for people to get to work and go about their daily lives.

The environmental effects of coastal recession also relate to the loss of land. An example of this is the National Trust area known as Golden Cap, close to the village of Seaton in Devon. The cliff there has receded 40 m in the past twenty years. This means that animals and birds are losing their breeding grounds, for example the soft cliff faces where sand martins breed.

Coastal recession at Happisburgh on the North Norfolk coast

The village of Happisburgh on the North Norfolk coast, which has a population of approximately 850, is one of the fastest eroding areas in the UK. The area was defended in 1958 with revetments – sloping structures designed to absorb the energy of the waves – which reduced the amount of erosion to about 50 cm a year. However, in 1995 the council stopped repairing the coastal defences, which caused the rate of erosion to accelerate. Since this time, 25 properties and the village's lifeboat launching station have been washed away. This has had a major impact on the people who live in the area (see Figure 2.15). The main area of concern is Beach Road, which terminates in the sea. Houses that were worth £80,000 when the coast was defended are now valued at a £1, even though their sea view improves each year! (See Figure 2.16).

Houses on Beach Road

Broken revetments

Remains of lifeboat station ramp

Happisburgh lifeboat station

Clay and sand cliff easily eroded

➡ **Figure 2.15** The lifeboat station and ramp, Happisburgh.

➡ **Figure 2.16** Beach Road terminating in the sea.

Coastal recession at Dawlish, Devon

The main railway line into the South West to Penzance runs along the coast at Dawlish. When the sea is rough the trains have to be cancelled or delayed. On one occasion 160 passengers were stranded for four hours when their train's electrics were affected by sea water that washed over the track. The train was eventually pulled into the station at Dawlish. The railway line is protected by a sea wall that was built in the 1800s; this wall has no wave-refracting curve and is in need of constant repair. The annual rebuilding and repair bill is £400,000. The situation was made worse in February 2014, when 80 m of sea wall beneath the railway collapsed in a storm and the railway line was left with no land beneath it. The storm also damaged the station and a long section of track (see Figure 2.17). The line was closed until early April. For two months commuters from the South West had to use a replacement bus service, which greatly lengthened their journeys. The houses behind the breach in the sea wall were also damaged. Although the residents were rescued, they did lose some of their possessions.

⬆ **Figure 2.17** The Dawlish railway line after the storm.

What are the effects of coastal flooding on people and the environment?

Around 2.1 million properties are at risk from flooding in England, with nearly 50 per cent of these being at risk from flooding from the sea. Some of the effects of coastal flooding on people are:

- damage to people's homes and belongings from water
- loss of life from drowning
- the contamination of fresh water supplies by sewage water
- bridges and roads can be washed away
- disruption to gas and electricity supplies.

The environmental effects of coastal flooding mainly concern the loss of land to the sea. In some areas of the country, land is being allowed to flood as part of the coastal defence technique known as **managed retreat**, where the land is flooded in a controlled way. This forms new habitats for wading birds but it does mean a loss of land for farmers. Another more dramatic type of coastal flooding is when the sea floods land due to storms, causing trees and vegetation to be washed away and crops lost due to inundation by sea water.

In December 2013 a high spring tide combined with an area of low pressure and strong northerly winds caused a storm surge that flooded much of the 45-mile coastline of North Norfolk. Following warnings given by the Environment Agency, flood wardens in the area were able to evacuate over 200 houses before the flood occurred. Although people were taken to safety, 152 houses and businesses were damaged as a result of the storm surge. There was also extensive damage to the sea defences in the area and many beach huts and chalets on the coast were damaged beyond repair. The public were warned that their drinking water might be contaminated and of other health risks such as rats in properties that had been flooded.

The Cove House Inn on the Isle of Portland is continually fighting back the waves when there is a storm surge. It lies behind the 9 m sea defences built at Chiswell, Portland. During one storm in 2014, 18 m waves were crashing over the top of the three-storey building, smashing windows with pebbles, and sea water was pouring into the bar and living accommodation. The building itself is made from Portland stone which is very strong, although it does bear the battle scars of the many pebbles which have been thrown against it. The road to Portland from the mainland was closed twice during the winter of 2014 but, in the future, it may not be defended and it will become a true island, cutting off the 25,000 people who live there from the mainland.

KEY TERMS

Hard engineering – this method of coastal management involves major construction work, for example sea walls.

Soft engineering – this method of coastal management works or attempts to work with the natural processes occurring on the coastline, for example beach nourishment. They tend to be visually unobtrusive and do not involve major construction work.

What are the main types of hard and soft engineering used on the coastline of the UK?

A number of different types of coastal defences are used to defend the UK coastline, but what are their advantages and disadvantages and how do they change coastal landscapes? Coastal defences can be classified as either hard or soft engineering techniques. For the purpose of this chapter the following defences will be discussed:

- hard engineering: sea walls, groynes, **rip rap**.
- soft engineering: **beach nourishment**, **offshore reefs**.

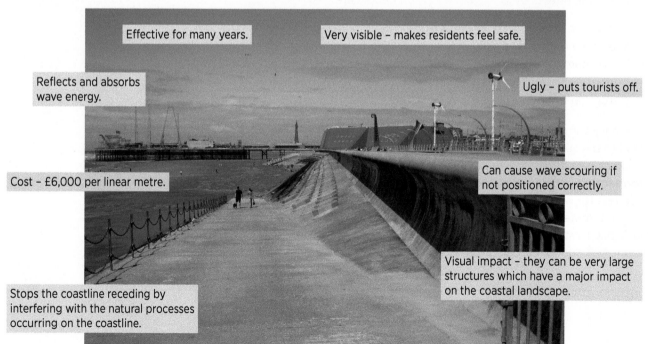

Effective for many years.

Very visible – makes residents feel safe.

Reflects and absorbs wave energy.

Ugly – puts tourists off.

Cost – £6,000 per linear metre.

Can cause wave scouring if not positioned correctly.

Visual impact – they can be very large structures which have a major impact on the coastal landscape.

Stops the coastline receding by interfering with the natural processes occurring on the coastline.

⬆ **Figure 2.18** Sea walls are usually made of concrete; the newer ones have a recurved top, like this one at Blackpool.

Cost – £400 per metre for 1 metre high wooden groyne.

Unattractive.

Effective for many years.

Keeps beach in place for the tourist industry.

Difficult to walk along the beach.

Prevents longshore drift – sand builds up on one side of the groyne.

Can have a visual impact.

Disrupts the natural processes at work on the beach.

⬆ **Figure 2.19** Groynes are usually made of rock or wood like these at St. Bees in Cumbria; they stretch from the coastline into the sea.

Can make beach inaccessible for tourists.

Effective for many years.

Not effective in storm conditions.

Dissipates wave energy.

Introduces foreign rock types to an area.

Unattractive.

Visually intrusive if placed on a sandy beach.

Can be cheap depending on rock type used.

⬆ **Figure 2.20** Rip rap are large rocks placed in front of a cliff; these are below a landscaped cliff at Whitby in Yorkshire.

Disrupts home owners – large noisy lorries regularly visit the area to replenish the beach.

Cheap – £6,500 per 100 metres.

Provides beach for tourists.

Looks natural.

May affect plant and animal life in the area.

Requires constant maintenance as it is washed away quickly.

The beach dissipates wave energy and is the best form of natural defence.

Good use of sand dredged from harbours and ports.

⬆ **Figure 2.21** Swanage beach before (2005) and after (2007) the placing of sand and pebbles on a beach, known as beach nourishment.

Cost – £1,950 per metre.

Difficult to install the reefs.

The waves break further offshore, which reduces their erosive power.

May be removed by heavy storms.

Visual impact – they change the way that the coastal landscape looks.

They allow the build up of sand due to the reduction in wave energy.

They interfere with natural processes such as longshore drift.

⬆ **Figure 2.22** Offshore reefs in Norfolk: enormous concrete blocks, natural boulders or even tyres are sunk offshore to alter wave direction and dissipate wave energy.

1 Describe one effect of coastal recession on people.
2 Using Figure 2.16 (see page 18), describe the effects of coastal flooding on people and the environment.
3 Name one soft and one hard engineering technique.
4 What is the difference between hard and soft engineering?
5 How does the building of offshore reefs lead to changes in the coastal landscape?

Extension

Explain the formation of the discordant coastline to the east of the Isle of Purbeck.

Review

By the end of this section you should be able to:

✓ understand how urbanisation, agriculture and industry have affected coastal landscapes
✓ recognise the effects that coastal recession and flooding have on people and the environment
✓ know the advantages and disadvantages of different coastal defences used on the coastline of the UK
✓ understand how coastal defences can lead to change in coastal landscapes.

Practise your skills

Draw a field sketch of Figure 2.24 (see page 24). Include the following on the sketch:

1 Four landforms created by erosion.
2 Two landforms created by deposition.
3 Name processes that occur on this coast.
4 Name the rock types (see Figure 2.9, page 13).
5 A class of students recorded the size of pebbles at two sites at each end of a beach. Their results are in the table.

Site 1 Pebble size (cm) at western end of beach	Site 2 Pebble size (cm) at eastern end of beach
12, 16, 10, 11, 6, 10, 8, 9, 6, 7, 14, 10, 15, 12	4, 9, 5, 3, 7, 5, 3, 4, 5, 3, 2, 7, 2, 5

a) Calculate the mean, mode and range for the pebble sizes in the table.
b) Draw out and complete the dispersion diagram, Figure 2.23, by plotting the information on pebble sizes for the eastern end of the beach. Mark on the dispersion diagram the median, the upper quartile, the lower quartile, and the interquartile range for both of the sites.
c) Compare the results for site 1 and site 2.
d) Suggest reasons for your results. Use your knowledge and understanding of beach processes to help you.

Suggested hypothesis: the landforms that make up the coastline of Swanage Bay are landforms resulting from deposition.

Primary fieldwork:

- draw field sketches of the landforms in Swanage Bay
- take photographs of the landforms in Swanage Bay and then anotate them.

Suggested hypothesis: the coastal processes at work in Swanage Bay have implications for people who live in the area.

Primary fieldwork:

- take measurements either side of selected groynes to measure the extent of longshore drift
- take photographs of cliff faces to show evidence of slumping and erosion.

Secondary fieldwork:

- use the internet to research recent cliff falls in the area.

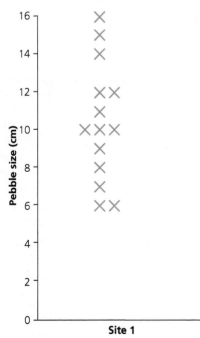

▲ **Figure 2.23** A dispersion diagram of pebble size.

Located example Distinctive coastal landscapes are the outcome of the interaction between physical and human processes

⬆ **Figure 2.24** An aerial photograph of part of the Isle of Purbeck.

The Isle of Purbeck in Dorset has a distinctive coastal landscape that formed due to the interaction of physical and human processes. The location is significant because it is part of the Jurassic Coast of Dorset, famous for fossils and coastal features such as headlands, bays and stacks. The Isle of Purbeck is made up of a concordant coastline to the south of the promontory and a discordant coastline to the east where the main settlement, Swanage, is located.

The coastal landscape formed due to the type and formation of rock present in the area: this is shown in Figure 2.26, a simplified geological map of the area. The coastline to the east of the Isle of Purbeck is made up of rocks with varying resistance to erosion that lie at right angles to the sea. This has allowed the sea to erode the rock types at different speeds, forming headlands and bays (see page 8 for information on the different types of erosion and Figure 2.9 on page 13).

Physical or human process	Impact on the coastal landscape
Coastal erosion	The headland of Ballard Down is constantly changing due to erosion and weathering. Originally there were two stacks off the coast – Old Harry and his wife – but, in 1896, Old Harry's wife collapsed forming a stump.
Landslips	The coastline to the south of Ballard Down has frequent landslips, causing the coastal path to have to be redirected on a number of occasions.
Coastal defences	In 2005–6 new coastal defences were built in Swanage Bay consisting of eighteen groynes and beach nourishment. This changed the area by creating a new higher beach, although it will have to be replenished every twenty years due to the erosion rates in the area.
Human development	The building of Swanage town, especially the houses and hotels on the cliff, have made the problem of land slipping in the area worse.
Tourism	Studland Bay to the north of the area is a tourist hot spot. The beach is owned and managed by the National Trust. The area is protected from excessive tourist damage by limiting the parking available and, therefore, the number of people who can access the beach. The sand dunes are also protected by being fenced off. In this way change to the area from human processes is being managed.

⬆ **Figure 2.25** Physical and human processes that have an impact on the coastal landscape around the Isle of Purbeck.

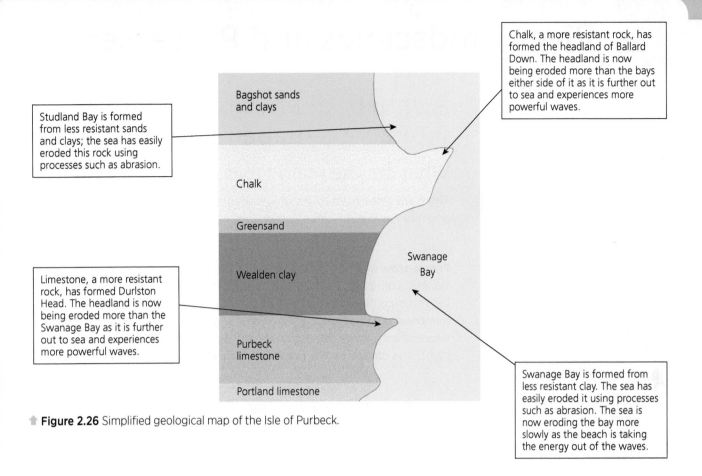

Chalk, a more resistant rock, has formed the headland of Ballard Down. The headland is now being eroded more than the bays either side of it as it is further out to sea and experiences more powerful waves.

Studland Bay is formed from less resistant sands and clays; the sea has easily eroded this rock using processes such as abrasion.

Bagshot sands and clays

Chalk

Greensand

Wealden clay

Swanage Bay

Limestone, a more resistant rock, has formed Durlston Head. The headland is now being eroded more than the Swanage Bay as it is further out to sea and experiences more powerful waves.

Purbeck limestone

Portland limestone

Swanage Bay is formed from less resistant clay. The sea has easily eroded it using processes such as abrasion. The sea is now eroding the bay more slowly as the beach is taking the energy out of the waves.

⬆ **Figure 2.26** Simplified geological map of the Isle of Purbeck.

Review

By the end of this section you should be able to:

✓ recognise the significance of the location of the Isle of Purbeck
✓ understand how physical processes formed the discordant coastline in the Swanage area
✓ describe and explain how physical and human processes have changed the coastal landscape of the area.

Practise your skills

1 Use the geology maps of the British Geological Survey (www.bgs.ac.uk/data/mapviewers/home.html) to link the shape of the coastline at Swanage to the geological formation.

Examination-style questions

1 Study Figure 2.13 on page 16. Identify **one** depositional landform shown on Figure 2.13. (1 mark)
2 Study Figure 2.4 on page 10. State **one** type of weathering which might have an impact on this landscape. (1 mark)
3 Groynes are an example of hard engineering.
   Explain **one** way that groynes help to protect coastal landscapes. (2 marks)
4 Examine how physical processes work together to form the stack shown in Figure 2.4. (8 marks)

Total: 12 marks

River Landscapes and Processes

To study the physical processes that interact to shape river landscapes.

Learning outcomes

▶ To understand the impact of weathering, mass movement and erosion on river landscapes.
▶ To understand the ways that rivers transport and deposit material.
▶ To recognise that rivers are different between their upper, mid and lower courses.
▶ To be able to explain why river characteristics change along the course of a river, such as the River Creedy.
▶ To know how the UK's weather and climate affect river processes and impact on landforms and landscapes.

A variety of physical processes interact to shape river landscapes

What are the main types of weathering?

There are three main forms of weathering: mechanical, chemical and biological.

Mechanical weathering

Freeze–thaw weathering, or frost action, is when water gets into cracks in rocks. When the temperature falls below freezing, the water will expand as it turns into ice. This expansion puts pressure on the rock around it and fragments of rock may break off. This type of weathering is common in highland areas where the temperature is above freezing during the day and below freezing during the night. This can cause the break-up of rocks on riverbanks.

Chemical weathering

Rainwater contains weak acids that can react with certain rock types. The carbonates in limestone, for example, are dissolved by these weak acids and this causes the rock to break up or disintegrate. This can be seen on limestone pavements or on limestone cliffs in a gorge. Figure 3.1 shows chemical weathering on the Devil's Pulpit rock in the Lower Wye Valley.

Biological weathering

This is the action of plants and animals on the land. Seeds that fall into cracks in rocks will start to grow when moisture is present. The roots the young plant puts out force their way into cracks and, in time, can break up rocks. The rocks can then be washed away by rainwater, leaving the roots exposed (see Figure 3.2 on p page 27). Burrowing animals, such as rabbits, can also be responsible for the further break-up of rocks.

◀ **Figure 3.1** Chemical weathering on the Devil's Pulpit, Lower Wye Valley.

⬆ **Figure 3.2** Biological weathering.

What is mass movement?

Mass movement is when material moves down a slope due to the pull of gravity. There are many types of mass movement but, for the purposes of this chapter, only slumping and sliding will be discussed. Slumping, also known as rotational slipping, is common on riverbanks and involves an area of riverbank slipping into the river. Due to the nature of the slip, it leaves behind a curved surface. This is very common on clay riverbanks. During dry weather the clay contracts and cracks; when it rains, the water runs into the cracks and is absorbed until the rock becomes saturated. This weakens the rock and, due to the pull of gravity, it slips down the slope on its slip plane.

How do rivers erode?

The water in a river is continually attacking its bed and banks in a process known as erosion. Many of the same processes are used by both the sea and rivers to erode land. The theory box should be referred to throughout this chapter to understand the processes of river erosion.

Processes of river erosion

Abrasion

Material that the river is carrying is thrown against the river bed and banks. These particles break off more rocks which, in turn, are thrown against the river bed and banks as the water continues to make its way downstream.

Hydraulic action

This is the pressure of the water being pushed against the riverbanks. It also includes the compression of air in cracks: as the water splashes against the riverbanks, it compresses air in cracks; this puts even more pressure on the cracks and pieces of rock may break off.

Solution

This is a chemical reaction between certain rock types and the minerals in the river water. This is particularly evident in limestone areas where rivers can eat away at the rock, disappearing underground for part of their course.

Attrition

This is a slightly different process that involves the wearing away of rocks that are in the river. Rocks in the river roll around; as they do so, they grind away at each other until smooth pebbles or sand are formed.

How do rivers transport materials?

Rivers can transport materials in a number of ways: traction, saltation, suspension and solution.

- **Traction** – large **sediment** such as pebbles roll along the river bed.
- **Saltation** – small pieces of shingle or large grains of sand are bounced along the river bed.
- **Suspension** – small particles such as sand and clays are carried in the water. This can make the water look cloudy, especially after heavy rainfall or when the river has lots of energy.
- **Solution** – some minerals are dissolved in river water. This can change the colour of the water due to the minerals that are present.

KEY TERMS

Gradient – the slope over which the river loses height.

↑ **Figure 3.3** How a river transports materials.

What is deposition?

Deposition is the laying down of materials, such as sand and pebbles, that are being transported by the river. Rivers deposit materials when they slow down and lose energy, such as on the shallow bank of a river.

How river landscapes contrast between the upper course, the mid course and the lower course

As a river flows downstream, the contrasts in the river landscapes that occur mostly relate to differences in the river's energy. When it is in an upland area, a river has the power to erode downwards, as it is way above sea level, and it forms a V shaped valley. As the gradient (slope) of a valley decreases, the river uses its energy to transport the material it has eroded. Due to the lack of gradient, it begins to erode laterally (sideways). As the river moves closer to sea level, the gradient decreases further. Although the river is still eroding sideways at this point, deposition is the most important process, and the valley becomes wider and flatter in the lower course. This change from erosion to deposition helps to explain the change in landforms and the shape of the river valley as the river moves towards the sea (see Figure 3.4 on page 29).

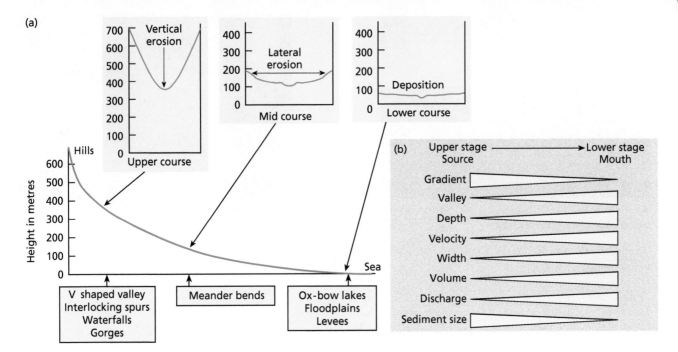

↟ **Figure 3.4** The features of a river's course.

The long profile of a river shows the steep gradient at the source gradually becoming more gentle until the river reaches sea level. These changes usually show a river to be split into three sections, known as the upper, mid and lower courses.

As a river moves downstream, its discharge also changes. Discharge is the amount of water passing a specific point at a given time and is measured in cubic metres per second. The discharge depends on the river's velocity and volume. The volume is the amount of water in the river and the velocity is the speed of the river. A river's discharge is equal to its velocity multiplied by its volume.

As a river moves towards the sea, its discharge will increase because of the increased volume as more tributaries join the river. The velocity of the river is determined by the amount of water that is touching the river's bed and banks. If the river is deeper, there will be less contact between the river and its banks and bed, therefore less friction will occur and the river velocity will be greater.

Why do river characteristics change along the course of the River Creedy?

The River Creedy flows through 16 km of mid Devon. Its source is in the hills to the north of Crediton. It ends at its confluence with the River Exe, which is just south of Cowley Bridge. In the upper course the river is shallow and narrow with a steep gradient. As it moves downstream it meets other rivers such as the Binneford. The river becomes wider and deeper as it gains more water from other rivers. To the southeast of Crediton it is met by the River Yeo, adding more water to the river. This gives the river more power to erode. The river is also moving away from the hills into flatter areas, so the gradient of the river is becoming less and the river is less likely to erode vertically and more likely to erode laterally, forming meander bends.

A
Upper course of the River Creedy

Characteristics of the upper course

Characteristic	Detail
Channel shape	Width 1 m, depth 10 cm
Valley profile	V shaped
Gradient	10%
Discharge	0.66 m³
Surface velocity	0.38 m per sec
Sediment size	10 cm
Sediment shape	More angular

Key

	Minor road
	A road
	Built up area
	Rivers
•10 m	Spot height

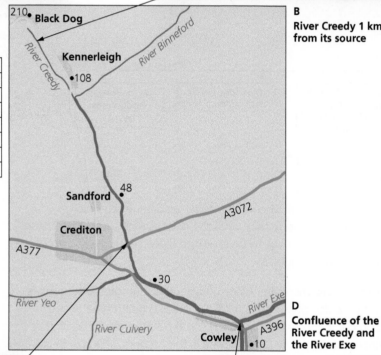

B
River Creedy 1 km from its source

D
Confluence of the River Creedy and the River Exe

C
Mid course of the River Creedy

Characteristics of the mid course

Characteristic	Detail
Channel shape	Width 10 m, depth 50 cm
Valley profile	Wide, flat-bottomed V shape
Gradient	0.5%
Discharge	2.8 m³
Surface velocity	0.56 m per sec
Sediment size	15 cm
Sediment shape	Between angular and rounded

Characteristics of the lower course

Characteristic	Detail
Channel shape	Width 15 m, depth 1.2 m
Valley profile	V shaped with a very wide, flat bottom
Gradient	0.1%
Discharge	15.5 m³
Surface velocity	0.86 m per sec
Sediment size	3 cm
Sediment shape	More rounded

⬆ **Figure 3.5** The course of the River Creedy.

How the UK's weather and climate affect river processes and impact on landforms and landscapes

The **seasonality** of the UK's weather and climate affect the rate at which a river can erode. In times of storms or heavy rainfall, rivers will contain a lot of water. This gives them greater erosive power and landforms such as river cliffs will be attacked by the water. During times of heavy rainfall trees on riverbanks can have their roots undermined by erosion and mass movement, eventually causing the bank to collapse. **Levees** are depositional landforms that form when the river is in flood or during times of heavy rainfall. When droughts occur rivers have less water and therefore less power to erode. In the winter the differences between day and night-time temperatures can cause freeze–thaw weathering on riverbanks, possibly causing the banks to collapse.

The human river landscape, such as river defences, are in need of repair more regularly due to the increasing regularity of storms in the UK. Many urban areas have had to install river defences, causing a change to the river landscape on the **floodplain**.

Review

By the end of this section you should be able to:

✓ explain the impact of weathering, mass movement and erosion on river landscapes
✓ describe the ways that rivers transport and deposit material
✓ describe the ways that rivers differ between their upper, mid and lower courses
✓ explain why river characteristics change along the course of a river, for example the River Creedy
✓ know how the UK's weather and climate affect river processes and impact on landforms and landscapes.

ACTIVITIES

1 Which of the following is not a way that a river erodes materials?
  abrasion   attrition   solution   traction
2 Describe two of the ways that a river transports materials.
3 What is the difference between a long profile and a valley profile of a river?
4 How do rivers differ between their source and their mouth? Refer to width, depth, gradient and discharge in your answer.
5 Explain how the UK's weather and climate affect river landscapes.

Extension

With reference to a river you have studied, explain the changes between its source and its mouth.

Practise your skills

A class of students recorded the size of sediment at two sites on a river – the source and close to the mouth. Their results are in the table.

Site 1 Sediment size (cm) at the source	Site 2 Sediment size (cm) at the mouth
11, 8, 10, 11, 6, 10, 8, 9, 9, 7, 8, 10, 8, 12.	4, 3, 5, 4, 7, 5, 3, 4, 5, 3, 2, 7, 2, 5.

1 Calculate the mean, mode and range for the pebble sizes in the table.
2 Draw out and complete a dispersion diagram similar to Figure 2.23 on page 23 by plotting the information on sediment sizes for sites 1 and 2. Mark on the dispersion diagram the median, the upper quartile, the lower quartile and the interquartile range for both of the sites.
3 Compare the results for site 1 and site 2.
4 Suggest reasons for your results. Use your knowledge and understanding of river processes to help you.

Fieldwork ideas

How do the characteristics of a river change as it flows downstream?

Primary fieldwork:

- measure width, depth, velocity, gradient, pebble size and shape at six locations, with two measurements to be taken in the upper course, two in the mid course and two in the lower course
- the measurements could be shared between groups but all students should use each of the techniques
- use the measurements for width, depth and velocity to calculate the discharge of the river.

River erosion and deposition interact with geology to create distinctive landforms within river landscapes

LEARNING OBJECTIVE

To study the distinctive landforms created when river erosion and deposition interact with the geology of an area.

Learning outcomes

▶ To be able to describe and explain the formation of landforms that are created by river erosion interacting with the geology of an area: interlocking spurs, waterfalls and gorges.
▶ To be able to describe and explain the formation of landforms that are created by river erosion and deposition: meanders and oxbow lakes.
▶ To be able to describe and explain the formation of landforms that are created by deposition: floodplains and levees.

What landforms are created by river erosion interacting with the geology of an area?

Most river valleys can be split into three courses: the upper course, the mid course and the lower course. A number of different landforms can be found in river valleys which are the result of the interaction of erosion and deposition with the geology of the area.

Interlocking spurs

The river in the upper course is shallow and a lot of the water is in contact with its bed and banks. There is a lot of friction. The main process occurring in this area is erosion. The gradient is usually steep; the river erodes downwards forming a V shaped valley. In Figure 3.6 the River Tees is in its upper course. It is winding its way between interlocking spurs of carboniferous limestone. The river has to go around the rocky outcrops because it does not have the power to go through them.

Waterfalls and gorges

A waterfall forms due to river erosion and the influence of the geology of the area. If a band of more resistant rock crosses a rivers course, a waterfall will form at this point (see Figure 3.8).

The less-resistant, softer rock is eroded more quickly, leaving an overhang of more-resistant, harder rock. In time the harder rock becomes too heavy and falls into the river below. Hydraulic action occurs as the water falls over the lip and splashes on to the back wall. Over time the waterfall moves back (retreats) up the valley forming a gorge. At the bottom of the waterfall a deep pool of water known as the plunge pool forms due to the power of the water falling into it and the process of abrasion – the rocks carried by the river erode the bottom and sides of the plunge pool. Figure 3.9 shows High Force waterfall on the River Tees in County Durham. There are a number of features which caused this waterfall and gorge to form. The area has a rainfall of approximately 800 mm spread evenly throughout the year. The position of the waterfall is where the River Tees crosses the Whin Sill. This hard igneous rock known as whinstone is resistant to the erosional processes of the river water. The lower section of the waterfall is made up of carboniferous limestone. This

⬇ **Figure 3.6** Interlocking spurs.

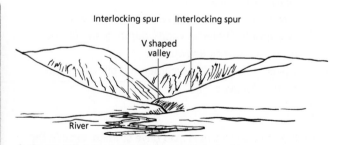

⬆ **Figure 3.7** A field sketch showing interlocking spurs.

is a less resistant rock than whinstone and therefore has eroded away at a faster rate. Between these two rocks is a layer of sandstone which is also less resistant then whinstone. This has caused the waterfall to slowly move upstream leaving a deep narrow gorge. The gorge is approximately 700 metres long and the drop of the fall is approximately 21 metres.

What landforms are created by river erosion and deposition?

Meander bends

These can be found on the river's course where the river is eroding and depositing material forming a bend in the river. The outside of a meander has the deepest water because this is where the greatest erosion takes place. The water is moving fastest at this point due to the lack of friction, eroding the bank using abrasion and forming a river cliff. Less water is in contact with the bed and banks because the river is deeper on this side.

A slip-off slope (point bar) forms on the inside of the meander bend because of deposition. Deposition occurs on the inside because the water is moving more slowly and is shallower. As a result, there is more friction here and the river is less powerful. The river is therefore unable to carry its load and deposition takes place. An underwater current takes some of the eroded material from the river cliff across the river and deposits it on the slip-off slope (see Figure 3.11).

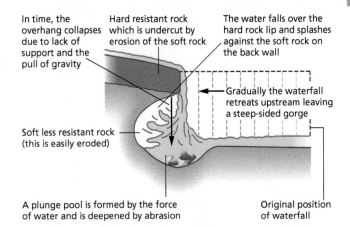

In time, the overhang collapses due to lack of support and the pull of gravity

Hard resistant rock which is undercut by erosion of the soft rock

The water falls over the hard rock lip and splashes against the soft rock on the back wall

Gradually the waterfall retreats upstream leaving a steep-sided gorge

Soft less resistant rock (this is easily eroded)

A plunge pool is formed by the force of water and is deepened by abrasion

Original position of waterfall

⬆ **Figure 3.8** The formation of a waterfall.

⬆ **Figure 3.9** High Force waterfall, County Durham.

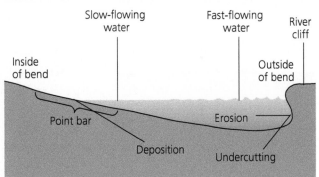

Slow-flowing water

Fast-flowing water

River cliff

Inside of bend

Outside of bend

Point bar

Erosion

Deposition

Undercutting

⬆ **Figure 3.11** A cross-section of a meander bend.

⬆ **Figure 3.10** Meander bend, Ardèche river, France.

Oxbow lakes

Meander bends can become very large. With continual erosion on the outside of the banks and deposition on the inside, the ends of the meander bend become closer (see Figure 3.12). When flooding occurs, the river is able to cut through the gap and, in time, forms a new straight channel. Continued deposition of alluvium at times of low flow results in the old bend of the river becoming cut off. This is called an oxbow lake.

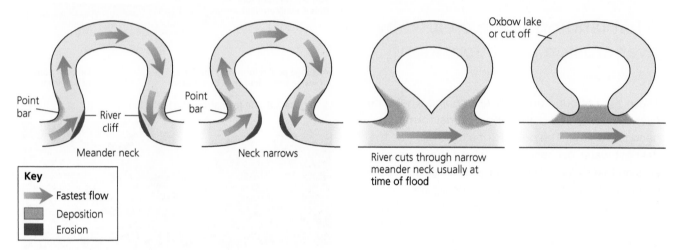

Key

→ Fastest flow

▨ Deposition

■ Erosion

⬆ **Figure 3.12** The formation of an oxbow lake.

What landforms are created by deposition?

Floodplains and levees

A floodplain is the low flat area of land on either side of a river. It can be found in the middle course of a river, but is more usually found in the lower course. It is formed by the migration of meanders downstream. Meanders are formed by lateral erosion which causes the bend to move across and down the valley in the direction of the river's flow. The outside of the bend, where erosion is greatest, moves the bend in that direction and the inside bend fills in the floodplain with the deposition that occurs there.

When the river contains too much water to stay within its channel, it floods the surrounding land. As it moves away from its channel, it becomes shallower and friction increases. The river has less energy and, therefore, must drop some of the load it is carrying. It drops the largest amount of material close to the river channel. After a number of floods, this builds up to form levees (see Figure 3.13). The river water drops the heaviest material first.

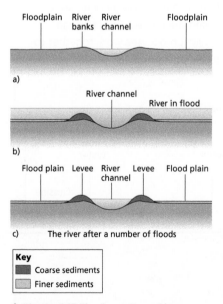

Key

■ Coarse sediments

▨ Finer sediments

⬆ **Figure 3.13** The formation of levees.

Review

By the end of this section you should be able to:

✓ describe and explain the formation of interlocking spurs, waterfalls and gorges

✓ describe and explain the formation of meanders and oxbow lakes

✓ describe and explain the formation of floodplains and levees.

Human activities can lead to changes in river landscapes that affect both people and the environment

■ LEARNING OBJECTIVE

To study how human activities can lead to changes in river landscapes that affect both people and the environment.

Learning outcomes

▶ To know how urbanisation has affected river landscapes.
▶ To recognise how agriculture has affected river landscapes.
▶ To identify how industry has affected river landscapes.
▶ To recognise the physical and human causes of river flooding.
▶ To recognise the effects of river flooding on people and the environment.
▶ To know the advantages and disadvantages of different flood defences used on rivers in the UK.
▶ To understand how river defences can lead to change in river landscapes.

KEY TERMS

Interception – when trees stop precipitation hitting the ground surface.

Throughflow – when water travels through soil towards a river.

Physical causes – any occurrence that is natural.

Human causes – any occurrence that is created by humans.

Urbanisation – the increase in the number of people living in towns and cities compared to the number of people living in the countryside.

How have human activities such as urbanisation, agriculture and industry affected river landscapes?

The building of settlements in river valleys and on river estuaries has had a major impact on the landscape. The settlements have a visual impact, an impact on wildlife, but also a major impact on the natural processes that are occurring in the area. Settlements developed on riverbanks originally for water supply and transport but, as the urban area spread, its impact has increased. The building of roads and drains means that water can reach the river more quickly after rainfall. This means that the river is much more likely to flood because of the disruption in the balance of water flow to the river. This, in turn, means that river defences are built to protect people's homes, which further disrupts the natural processes at work. For example, the river is **channelised**, which means that the original bank is reinforced. This stops the natural formation of meanders and speeds up the flow of the water. The building of large ports for industry on many river estuaries has also disrupted the natural processes at work in the river. For example, Poole Harbour in Dorset is regularly dredged of river sediments to stop it silting up so that large container boats can continue to access the harbour.

Agriculture has also had a number of impacts on river valleys. Many trees have been felled to make way for farming land, which interferes with the process of interception and has an impact on throughflow. Fewer trees means that rainwater will reach the river faster, impacting on erosion rates in the river because it will be more powerful. Farmers are also using more chemical fertilisers, which have an impact on the ecology of the river, causing algae bloom and water channels to become clogged with vegetation. This, in time, will impact on the river landscapes, as is particularly evident in Norfolk (see Figure 3.14). Farmers also drain land close to rivers with artificial drainage ditches that allow water to flow to the rivers of the area more quickly. This can cause flooding as the natural river landscape has been changed.

⬆ **Figure 3.14** Algae bloom in a stream in Norfolk.

What are the physical and human causes of river flooding?

Rivers overflow their banks or flood when there is more water available than their channel can hold. There are both physical and human causes of flooding.

What are the effects of river flooding on people and the environment?

Around 2.1 million properties in England are at risk from flooding, with just over 50 per cent of them at risk from flooding from rivers. Some of the effects of river flooding on people are:

- damage to people's homes and belongings from water
- loss of life from drowning of people and livestock
- contamination of fresh water supplies by sewage water
- communication links can be destroyed if bridges and roads are washed away
- disruption to gas and electricity supplies
- fields of food crops can be flooded, so there could be a lack of food in the area.

The destruction of crops is a short-term effect of river flooding on the environment, but it can take the land a number of years to recover from being covered by water. It can also cause different plant species to develop on the flooded land, making it expensive for farmers to get the land ready to be cultivated again. The ecosystem of the area is also affected with animals such as rabbits being drowned and natural vegetation that these animals feed on being lost. It can take many years for the natural environment to recover from the effects of a river flood.

Physical causes	Human causes
Heavy rainfall: if there are large amounts of rain day after day, the water will saturate the ground and flow more quickly into the river.	Removal of vegetation on valley slopes: if there is less interception that water will move to the river more quickly.
Cloudburst in a thunderstorm: the rain droplets are so large and fall so quickly that there is no time for the water to sink into the ground; water runs very quickly into the river and causes flooding.	Settlements built on the floodplain: storm drains allow water to move into rivers at a greater speed and so make flooding more likely.
Sudden rise in temperature: a rapid thaw can happen; rivers are unable to cope with the amount of water and flood.	Global warming: melting of polar ice caps and a rise in sea levels, flooding low-lying coastal areas.
Silted up river channels: this makes the channel smaller and more likely to flood.	Dams may burst: which causes excess water in river channels and flooding of large areas.

⬆ **Figure 3.15** The physical and human causes of river flooding.

River flooding on the Somerset Levels and Moors

During the winter of 2013–14 the Somerset Levels suffered severe flooding. The Levels are an area of low-lying land with the highest points being only 3–4 m above sea level; much of the land is below sea level. They are dissected by many rivers, the largest being the Parrett and the Tone. The heavy rainfall started in December and didn't stop until February, leading to the flooding of over 600 houses and 11,500 hectares of farming land, including North Curry and Hay Moors and the Greylake area. It could take up to two years for the land to recover. Villages such as Thorney were abandoned, and Mulchelney and Moorland were cut off. The A361 between East Lyng and Burrowbridge was closed for almost three months, leaving residents with the only options of evacuating their homes or using boats to get their shopping.

⬆ **Figure 3.16** The flooded village of Moorland, Somerset.

Hard engineering – this method of river management involves major construction work, for example dams.

Soft engineering – this method of river management works or attempts to work with the natural processes occurring. They tend to be visually unobtrusive and do not tend to involve major construction work, for example washlands.

Hydrograph – a graph showing rainfall and river discharge over a specific period of time.

What are the main types of hard and soft engineering used on UK rivers?

A number of different types of defences are used to manage rivers, but what are their advantages and disadvantages and how do they change river landscapes? River defences can be classified as either **hard** or **soft engineering** techniques.

For the purpose of this chapter the following defences will be discussed:

● hard engineering: **dams, reservoirs, channelisation**

● soft engineering: floodplain zoning, washlands.

Practise your skills

1 Use the internet to find out the daily rainfall totals for January 2014 for Somerset and the river discharge for the River Parrett, Somerset. Use the figures to draw a **hydrograph** for this period.

2 Use the Environment Agency website to analyse the flood hazard in the area in which you live.

Dams – a large, usually concrete, structure built across a river valley to hold back water.

Very visible – makes residents feel safe.

Effective for many years.

They can be seen as ugly; may put tourists off.

Stores water behind the dam; hydroelectric power can be produced as water is released through the dam.

Can cause problems with salmon trying to make their way upstream.

Dams are very expensive to build.

They visually impact on river landscapes as the concrete they are made from is very unnatural.

Sediments can be trapped behind the dam causing a lack of deposition further downstream.

⬆ **Figure 3.17** The Derwent Dam and Reservoir in Yorkshire.

Reservoirs – large areas of water that are created after the flow of a river has been controlled, often by building a dam.

Very expensive to control the flow of the river.

Reservoirs can provide drinking water for urban areas.

Settlements and farming land can be lost when a valley is flooded.

Creates a large area of water that can be used for recreational activities.

River landscapes are flooded with loss of land and settlements.

Disrupts the natural processes at work in the river valley as sediment is trapped in the reservoir.

Has a visual impact on the river landscape as a large body of water is created.

↑ **Figure 3.18** The Derwent Reservoir, Yorkshire.

The water can travel faster to places downstream and possibly cause flooding there.

Channelisation – the river channel is made deeper, wider and straighter.

The river channel can hold more water, so less likely to flood.

Expensive.

Long lasting.

Visual, so makes residents feel safe.

River landscape is changed and can look unnatural.

Disrupts the natural processes at work in the river channel.

↑ **Figure 3.19** Channelisation in Reading.

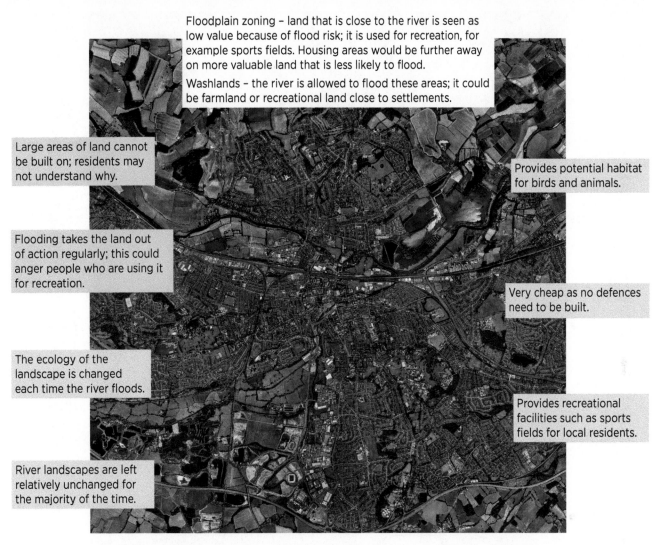

Floodplain zoning – land that is close to the river is seen as low value because of flood risk; it is used for recreation, for example sports fields. Housing areas would be further away on more valuable land that is less likely to flood.

Washlands – the river is allowed to flood these areas; it could be farmland or recreational land close to settlements.

Large areas of land cannot be built on; residents may not understand why.

Provides potential habitat for birds and animals.

Flooding takes the land out of action regularly; this could anger people who are using it for recreation.

Very cheap as no defences need to be built.

The ecology of the landscape is changed each time the river floods.

Provides recreational facilities such as sports fields for local residents.

River landscapes are left relatively unchanged for the majority of the time.

⬆ **Figure 3.20** Reading town centre: the River Thames has undeveloped land along most of its course through the town; this allows the river to flood without causing damage to people's homes and businesses.

Review

By the end of this section you should be able to:

✓ understand how urbanisation, agriculture and industry have affected river landscapes
✓ recognise the physical and human causes of river flooding
✓ recognise the effects that river flooding have on people and the environment
✓ know the advantages and disadvantages of river management schemes
✓ understand how river management schemes can lead to changed river landscapes.

ACTIVITIES

1 Describe one effect of urbanisation on river landscapes.
2 Looking at Figure 3.16 on page 36, describe the effects of river flooding on people and the environment.
3 Name one soft and one hard engineering technique.
4 What is the difference between hard and soft engineering?
5 How does the building of reservoirs lead to changes in river landscapes?

Extension

Evaluate the costs and benefits of hard and soft engineering techniques on river landscapes.

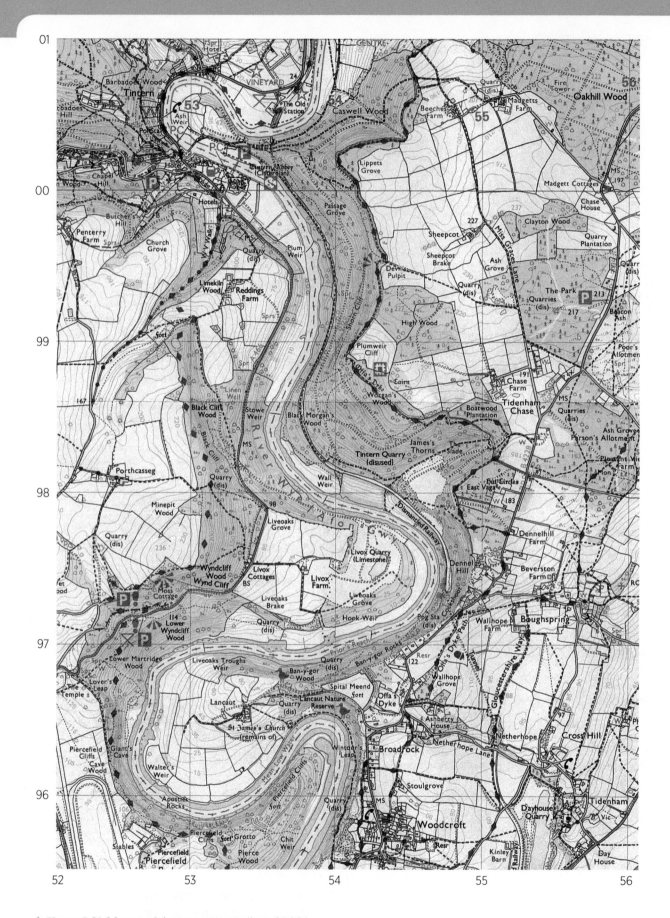

Figure 3.21 OS map of the Lower Wye Valley, 1:25 000.

Distinctive river landscapes are the outcome of the interaction between physical and human processes

To study how the interaction between physical and human processes produces distinctive river landscapes.

Learning outcomes

▶ To know the significance of the location of the Lower Wye Valley between Chepstow and Tintern.
▶ To know how physical processes formed the distinctive river landscape in this section of the Wye Valley.
▶ To recognise how physical and human processes have changed the river landscape of the area.

The significance of the location of the Lower Wye Valley between Tintern and Chepstow is that it forms part of the border between England and Wales. It is also significant because of its spectacular scenery, which could have only formed in this particular location due to the geology of the area. The location is also significant because it was where tourism was supposed to have started due to the beauty of the area. Many of the viewpoints that exist today were built for tourists in the mid-eighteenth century.

The formation of the Lower Wye Valley has caused disagreement between geologists because the course of the river today does not relate to the present relief. The landscape either formed on rocks that were more recent than the ones shown today, but these rocks have since been removed by erosion; or the drainage pattern we see today has been superimposed on the present relief. Alternatively, the landscape was formed by glacial melt water which was very powerful and able to cut its way through the limestone, forming the gorge that is there today. The River Wye flows over Old Red Sandstone as far as Tintern where the physical processes of the river have cut deep meanders. After this point the river flows through a gorge cut through carboniferous limestone.

⬇ **Figure 3.22** Physical processes at work in the Lower Wye Valley.

Physical or human process	Impact on (change to) the river landscape
Industry	Quarrying – the sides of the gorge have been extensively quarried for limestone for building materials and limekilns. This has increased the slopes of the gorge.

Iron ore smelting – the valley had a plentiful supply of water, iron ore and wood for charcoal; it was the perfect setting for early iron smelting in Britain. |
| River erosion | The river erodes and deposits material forming meanders and floodplains. See pages 33–35 for detail on these processes. |
| Weathering | The processes of mechanical, chemical and biological weathering are all present in the area, providing material for the river to use in erosion and deposition processes. |
| Forestry | Many trees were felled in the eighteenth and nineteenth centuries for shipbuilding and other industrial uses, such as making charcoal.

Up to the Second World War the woodlands were mainly deciduous. After this time extensive planting led to the area having 40 per cent of its woods either dominated by conifers or a substantial amount of conifers and a few broadleaf trees. Since the 1980s this planting has stopped and broad-leaved trees are now the main type being planted. Some woodlands were destroyed completely; others have appeared, such as on Coppet Hill where a wood has replaced previous open common pasture. |
| Human development | A road was built along the valley in the early nineteenth century and the railway followed in 1876. Before this the river was the economic backbone of the area allowing access for industry and tourists. Settlement in the valley goes back 12,000 years. Offa's Dyke, on the east bank of the river, was built in the eighth century. |
| Tourism | The Wye Valley was one of the earliest tourist honeypots with visitors flocking to the area in the 1700s. The cliff ascent and walks at Piercefield Park were landscaped at this time. Tourists still flock to the area. There are many lookout points, walks, a number of castles and Tintern Abbey, which dates back to the eleventh century. |

⬆ **Figure 3.23** Physical and human processes that have shaped the landscape of the Lower Wye Valley.

⬆ **Figure 3.24** The River Wye floodplain at Tintern.

Practise your skills

1 Use the geology maps of the British Geological Survey (www.bgs.ac.uk/data/mapViewers/home.html) to complete your own geological map of the Lower Wye Valley.
2 On the Digimap website (http://digimap.edina.ac.uk) find past and present maps of the Lower Wye Valley to investigate the changes to the human landscape, for example the removal of woodlands and the presence of quarries in the area.
3 Study Figure 3.21, the OS map of the Lower Wye Valley on page 40. Draw a sketch map of the area.

ACTIVITIES

1 What is the significance of the location of the River Wye landscape?
2 Describe how physical processes formed the Lower Wye Valley.
3 Use the information in Figure 3.23 to list the physical and human processes that have changed the River Wye landscape.
4 Explain how physical processes have changed the river landscape of the Lower Wye Valley.

Extension

Copy and complete the table below for the river landforms that can be seen on the OS map of the Lower Wye Valley in Figure 3.21 on page 40.

River landform	How it was formed

Review

By the end of this section you should be able to:

✓ recognise the significance of the location of the Lower Wye Valley between Chepstow and Tintern.
✓ know how physical processes formed the distinctive river landscape in this section of the Wye Valley.
✓ describe and explain how physical and human processes have changed the river landscape of the area.

Practise your skills

You will need to be able to identify certain settlement characteristics on OS maps. The site of a settlement is the land on which the settlement is built. The situation of a settlement is where the settlement is located in relation to other human and physical features in the area. The shape is the form that the settlement takes. Dispersed settlements are where individual buildings are spread out over an area, where there is no obvious centre to the village. Linear settlements have buildings either side of a main road. Nucleated settlements have buildings grouped closely together. They often form at crossroads or around village greens.

Use the OS map of the Lower Wye Valley on page 40 to help with the following questions.

1 Figure 3.24 was taken at grid reference 543995. In which direction was the camera pointing?
2 Give grid references for two tourist features found on the OS map.
3 Describe the shape of the settlement of Boughspring in grid square 5597.
4 Describe the site of the settlement of Tintern Parva in grid squares 5200 and 5300.
5 Describe the physical features of the land in grid square 5300.

Examination-style questions

1 Study Figure 3.22 on page 41. Identify **one** river landform shown in the photograph. (1 mark)
2 Study Figure 3.6 on page 32. State **one** type of weathering which might have had an impact on this river landscape. (1 mark)
3 Dams are an example of hard engineering.
   Explain **one** way that dams help to protect river landscapes. (2 marks)
4 Examine how physical processes work together to form the meander bend shown in Figure 3.10 on page 33. (8 marks)

Total: 12 marks

4 Glaciated Upland Landscapes and Processes

LEARNING OBJECTIVE

To study the physical processes that once operated in upland glaciated landscapes.

Learning outcomes

▶ To understand how glaciers erode.
▶ To understand the ways that glaciers transport and deposit material.
▶ To recognise that physical processes are still having an impact on upland glaciated areas.
▶ To know how past climate and current UK weather and climate affect processes that impact on upland glaciated landscapes.

A variety of physical processes interact to shape upland glaciated landscapes

What are the main types of glacial erosion?

Freeze–thaw (or frost) action occurs in glaciated areas when water gets into cracks in rocks. When the temperature falls below freezing, the water in the cracks will expand as it turns into ice. This expansion puts pressure on the rock around it and fragments of rock may break off.

A small crack in a rock fills with water during the daytime. As the water begins to freeze at night, it starts at the top, sealing the crack.

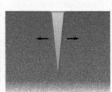
As the water freezes completely, its 9% growth exerts an outward force on the sides of the crack, increasing the size of the crack by a maximum of 9%.

If the ice thaws the next day the resulting water will not fill the crack, which is now both wider and deeper because of its 9% expansion. Dew or rainfall on the rock surface can refill the crack.

The process begins again, this time with a larger initial crack.

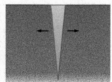
Again the crack expands by as much as 9%. Continued freezing and thawing, particularly with the daily addition of the water to keep the crack full, eventually leads to significant fracturing of the rock.

⬆ **Figure 4.1** The process of freeze–thaw.

Plucking is a form of glacial erosion. As a glacier moves down a valley it puts pressure on the valley sides and bottom. This pressure creates heat. (If you push your hands together, they will become warm because of the pressure you are using.) The heat causes a small amount of ice to melt, the water runs into cracks in the valley sides or bottom and almost immediately refreezes because of the cold temperatures. As the glacier moves, it then pulls away some of the rock face.

The rocks that have been removed from the valley sides and bottom by plucking (and freeze–thaw action) are carried in the ice of the glacier. As they move down the valley, they wear away (erode) the valley sides and bottom, causing more rock to be broken off. This process is another form of glacial erosion known as **abrasion**.

How do glaciers transport materials?

Glaciers can transport (move) large amounts of material. Imagine a large block of ice with bits of pebbles and rocks in it. The rock **debris** is moved in three ways. A glacier carries material on its surface at the sides close to the valley slopes, called the lateral moraine, and on the surface in the middle, known as the medial moraine. The glacier also has material in the ice itself. This material could have been on the surface of the glacier but over time was buried by snow falls.

How do glaciers deposit materials?

A glacier can deposit material in two ways. Material can be deposited by the glacier itself, such as moraines, which occurs due to a change in the power of the glacier. Where the glacier is in contact with the valley sides the ice moves more slowly due to friction between the ice and the valley. This causes the ice to deposit (lay down) some of the material it is carrying in a landform known as a lateral moraine. Glaciers also deposit material in glacial streams, either under the glacier or flowing out of the glacier. Glacial deposition in the **glaciated landscape** today can be seen in Figure 4.2 as a large mound covered in vegetation.

What processes are occurring in upland glacial landscapes in the UK today?

One of the processes occurring in glaciated landscapes is mechanical weathering or freeze–thaw action. This is when water gets into cracks in rocks. When the temperature falls below freezing, the water will expand as it turns into ice. This expansion puts pressure on the rock around it and fragments of rock may break off. This type of weathering is common in highland areas where the temperature is above freezing during the day and below freezing during the night.

Another process occurring in glaciated areas today is **mass movement**. This is when material moves down a slope due to the pull of gravity. This includes soil movement such as slumping (see page 9) and soil creep. This is when gravity very slowly pulls the water that is in soil down a slope; the soil moves slowly down the slope with the water. Unlike slumping it's not possible to see this happening, although the slope may appear rippled (like sheep paths around a hill). Mass movement also includes rock falls, which occur often in upland glaciated areas (see Figure 4.3).

KEY TERMS

Supraglacial debris – rock material that is carried on the surface of a glacier.

Englacial debris – rock material that is carried in the main body of the glacier.

Drift – all sediments deposited by glaciers.

Till – all the material deposited directly by a glacier, which is collectively known as moraine.

Fluvio-glacial material – rocks and other debris deposited by melting ice or glacial streams.

⬆ **Figure 4.2** Glacial deposit, Glen Catacol, Arran.

⬆ **Figure 4.3** Mass movement – a rock fall in Snowdonia.

KEY TERMS

Precipitation – any form of moisture that reaches the Earth's surface.

Diurnal – daily changes.

Seasonality – a pattern of change in the UK's weather between spring, summer autumn and winter.

How the past climate and current UK weather and climate affect processes that impact on upland glaciated landscapes

The past climate of the UK had a major impact on upland glaciated landscapes. In fact, it created them! They were formed during the last ice age, about 20,000 years ago. It is very difficult to imagine the UK covered in ice, but it did not occur overnight; the climate change happened gradually. As the climate got colder more precipitation fell as snow. The climate continued to get colder until the snow did not melt and, over a number of years, became compacted into ice, over a kilometre thick in Scotland! This is known as the ice age, and it was under these climatic conditions when an area of the UK north of the Tees/Exe line became permanently covered in ice that many of the features referred to in the next part of this chapter were formed.

The seasonality of the UK's present weather and climate affect the processes at work in glaciated areas. In winter, snow accumulates in mountainous areas such as Snowdonia but not enough to greatly change the landscape. Another process at work on the landscape is mechanical weathering, which will be more significant if there are large changes between day and night temperatures. The landscape is also constantly being changed by rivers and the sea. The way that these agents of erosion and deposition affect the landscape is covered in Chapter 2 (see pages 8–25) and Chapter 3 (see pages 26–43). Figure 4.2 shows a river eroding a terminal moraine on Arran, Scotland.

ACTIVITIES

1 Which of the following processes do not occur in glaciated areas?
      plucking   abrasion   deposition   attrition
2 Draw a set of diagrams to explain the process of freeze–thaw weathering.
3 Describe the process known as plucking.
4 How did the past climate of the UK have an impact on upland glaciated landscapes?

Extension

Research an upland glaciated area of the UK to investigate the impact of current weather and climate on upland glaciated landscapes.

Review

By the end of this section you should be able to:

✓ explain how glaciers erode
✓ describe the ways that glaciers transport and deposit material
✓ recognise that physical processes are still having an impact on upland glaciated areas
✓ know how past climate and current UK weather and climate affect processes that impact on upland glaciated landscapes.

Glacial erosion and deposition create distinctive landforms within upland glaciated landscapes

What landforms are created by glacial erosion?

A number of landforms are created by glacial erosion. For the purposes of this chapter they include: corries, arêtes, glacial troughs, truncated spurs, hanging valleys and roche moutonnées.

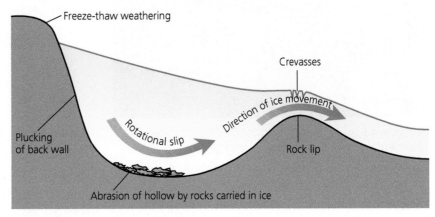

Freeze-thaw weathering

Crevasses

Direction of ice movement

Rotational slip

Plucking of back wall

Rock lip

Abrasion of hollow by rocks carried in ice

⬆ **Figure 4.4** The formation of a corrie.

Corries and arêtes

Corries are armchair-shaped hollows. They have a steep, rocky back wall, often up to 200 m high in the UK but much higher in the Alps. The corrie is fringed by sharp ridges, known as **arêtes**. A corrie begins to form when the snow that accumulates on high mountain slopes is compacted into ice. After a time, the ice starts to move due to the pull of gravity. As it moves, it carves out a corrie, usually in an area that already has a small depression. The ice erodes by plucking and forms the steep back wall; the hollow is formed by abrasion.

The rotational movement of the ice causes greater pressure at the bottom of the steep back wall and in the base of the hollow than it does at the front where the ice leaves the corrie. Because there is less pressure, erosion rates at the front of the hollow are slower. A rock lip forms here which can be increased in size by deposition of moraine. This works as a dam after glaciation, and a glacial (corrie) lake or **tarn** forms behind it. Freeze–thaw weathering occurs at the top of the steep back wall and adds rocks to the ice. These become embedded and are used in the process of abrasion.

If two corries form next to each other on a mountainside, their sides will be both steep and rocky. The piece of land between them will be a sharp ridge of rock (arête) that is continually attacked by freeze–thaw weathering and plucking.

Top of mountain

rie

Corrie

Original shape of the mountain

Arête

Original shape of the mountain

Pyramidal peak

rie

Corrie

Arête

◀ **Figure 4.5** The formation of arêtes and pyramidal peaks.

↑ **Figure 4.6** An aerial photograph of Blea Water, a glacial lake/tarn.

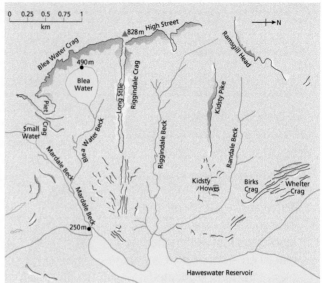

↑ **Figure 4.7** A map of the Blea Water area, Cumbria.

Blea water is a corrie lake (tarn) in Cumbria (Figure 4.6 and Figure 4.7). The characteristics of the landscape which caused the formation of the tarn were the fact that the area had been glaciated during the last ice age which left armchair shaped hollows for the tarn to form in. Blea water is on a north-east facing hillside. This helped the formation of corries because that aspect receives the least heat from the Sun. The sides of the corrie are steep and rocky forming arêtes. One of which is called Piet Crag which separates Blea water from Small water to its south east.

Practise your skills

Use Figures 4.6 and 4.7 to help you to answer the following questions.

1 Name two corrie lakes.
2 Name one arête.
3 Work out the difference in height between:
   a) The spot height at Blea Water and the pyramidal peak (top of the mountain)
   b) The spot height at Haweswater Reservoir and Blea Water.
   c) What is the total difference in height?
4 Study the aerial photo of the Blea Water area. Draw a sketch of the area including the corrie, two arêtes, the pyramidal peak, the steep back wall, the corrie lip and Blea Water.

Glacial troughs and truncated spurs

A **glacial trough** or U-shaped valley is formed by a valley glacier. A valley glacier will completely fill a valley in an upland area. By doing this, it has far more power to erode the whole valley than the original river, which only flowed across the valley floor. As a glacier moves down a valley, its immense power erodes any rock in its path. It does not need to go around crops of harder rock but simply removes them due to its immense power. In this way, interlocking spurs created by the original river are cut back to produce **truncated spurs** (see Figure 4.9 on page 49). The valley, which used to be V shaped with a river in the bottom, now has steep walls of bare rock for its sides and a flat bottom. It is also straighter and wider than it was before glaciation

The glacial trough in Figure 4.8 on page 49 formed in this valley because of the features of this particular area. The main valley had a number of streams which joined it. During the last ice age this meant that the ice in the main valley was very powerful. This enabled the glacier to cut a deep trough where there was once a V shaped river valley. Glacial troughs such as this are common in valleys in the French Alps.

↑ **Figure 4.8** A glacial trough in France.

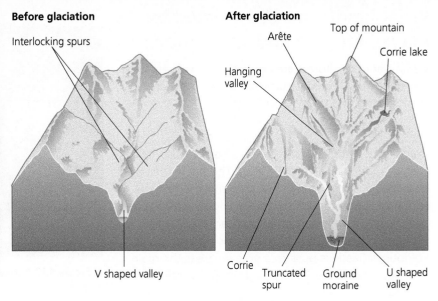

↑ **Figure 4.9** The formation of glacial troughs and hanging valleys.

Hanging valleys

The main valley glacier is very powerful due to the amount of ice that is being moved and the pressure that it exerts on the sides and bottom of the valley. Smaller tributary streams that join the main valley do not contain as much ice and have less power to erode. For this reason, they are not cut down as much as the main valley. After the ice has melted, these tributary valleys remain high on the sides of the main valley. The streams that flow in them now reach the main river by a waterfall down the steep sides of the main valley.

Roche moutonnées

These range from small rocky outcrops a few metres high to small hills a few hundred metres high. **Roche moutonnées** are formed from an outcrop of more resistant rock that was in the path of the ice as it moved across the upland area. The ice slows as it moves up the hard rock and more melting occurs. Abrasion will take place on this **stoss end** side, causing a smooth upstream slope to form. The ice is under less pressure on the lee side and refreezes around the rock, pulling rock fragments away as it moves. The main process here is plucking, forming a steeper slope.

What landforms are created by glacial deposition?

A number of landforms are created by glacial deposition. For the purposes of this chapter they include ground and terminal moraines.

Ground and terminal moraines

Moraine is material that is transported and deposited by a glacier. The glacier collects moraine through the processes of freeze–thaw action, abrasion and plucking. **Ground moraine** is deposited beneath the glacier and forms the flat valley floor. When the glacier reaches warmer temperatures it can no longer move down the valley, at which point it can no longer carry its load. The moraine deposited here is called **terminal moraine**.

> ### Practise your skills
>
> Study Figure 4.8 then draw an annotated sketch of the valley. Include in your sketch a hanging valley, a waterfall, a truncated spur, a corrie and an arête.

KEY TERMS

Pyramidal peak – the sharpened top of a mountain formed from the back walls of three or more corries.

Stoss end – the side of a landform facing the direction of the ice flow.

Volcanic plug – a volcanic cone from an extinct volcano.

Lee slope – the side of a landform facing away from the direction of the ice flow.

What landforms are created by glacial erosion and deposition?

A number of landforms are created by glacial erosion and deposition. For the purposes of this chapter they include: crag and tail, and drumlins.

Crag and tail

Crags form when a glacier meets a hard, resistant rock such as granite or a **volcanic plug**. The glacier will pass over the hard rock but leave it standing, eroding the softer rock in front of it. On the **lee slope**, material is deposited. This appears as a taper ramp away from the crag in the direction of the ice movement. Edinburgh Castle is built on a crag whereas the main street, known as the Royal Mile, is on the gently sloping tail.

⬆ **Figure 4.10** Abbey Craig near Stirling, an example of a crag and tail.

Drumlins

Drumlins are elongated landforms that are formed of till; they can be a kilometre or more in length, 500 m in width and 50 m in height. The slope facing the ice flow (the stoss end) is steeper than the slope that tapers away to ground level (the lee slope). Drumlins are often found in groups, called a swarm. Drumlins were formed when the ice was carrying a lot of material. If the ice had to slow down due to an obstruction in its path, it would deposit some of the material it was carrying. Due to the change in speed, most material would be deposited at this point and then gradually less as the ice continued on its path. This explains the shape of a drumlin, with its steep slope and then a gradual tapering away of the landform.

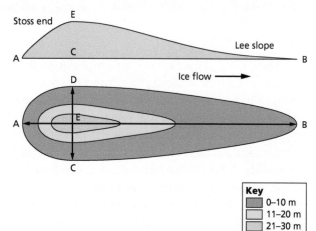

Key	
	0–10 m
	11–20 m
	21–30 m

⬆ **Figure 4.11** Plan and profile diagrams of a typical drumlin.

ACTIVITIES

1 Study the table below; it shows different glacial features and their descriptions. Match each feature with its correct description.

Feature	Description
Drumlin	An armchair-shaped hollow at the top of a hill.
Moraine	Smaller tributary valleys that are left high up on the sides of the main valley.
Corrie	Material carried by a glacier.
Hanging valley	A hill made of glacial deposits.

2 Explain how a corrie is formed.
3 Examine how physical processes work together to form the crag and tail shown on Figure 4.10.

Extension

1 How is moraine used to erode a valley?
2 How was the valley shown in Figure 4.8 formed?

Review

By the end of this section you should be able to:

✓ describe and explain the formation of truncated spurs, corries, glacial troughs, arêtes, hanging valleys and roche moutonnées
✓ describe and explain the formation of ground and terminal moraines
✓ describe and explain the formation of crag and tail, and drumlins.

Human activities can lead to changes in upland glaciated landscapes

How have human activities such as farming, forestry and settlement affected upland glaciated landscapes?

Upland glaciated landscapes are very rugged areas in which few people live. Settlements developed originally as farming communities and market towns but there has now been further development as tourism in upland areas continues to grow. Skiing resorts have been built in mountainous areas such as the Alps, but also in the UK in the Cairngorms in Scotland. The visual impact of these settlements can be seen on Figure 4.12; this is a major change to upland areas. Development has also had an impact on the ecosystems in upland areas; some animals and plants may become rare due to the building of settlements. The settlements also mean that more people are in the area, for example many tourists visit glaciated areas in the summer for walking holidays.

Farming has had an impact on glaciated areas. Originally many of the glaciated areas in the UK would have been woodland but they are now moorlands with rough pasture. For many years sheep farming was the main use of the land but this is changing with the demand for venison (deer meat). There will soon be more deer than sheep on the Island of Arran in Scotland because farmer are able to earn more money from game shooting and venison.

Forestry has also had an impact on upland areas. As mentioned earlier, much of the UK upland areas were once woodland; over time, it has been felled for fire wood, shipbuilding and many other uses. In the latter half of the twentieth century, native woodland was replaced with fast-growing evergreens. Now, however, forestry is much more in keeping with the landscape, being a mix of **coniferous** and **deciduous** woodland.

⬆ **Figure 4.12** Development of settlement in the French Alps due to tourism.

⬆ **Figure 4.13** Coniferous forest plantation in Snowdonia.

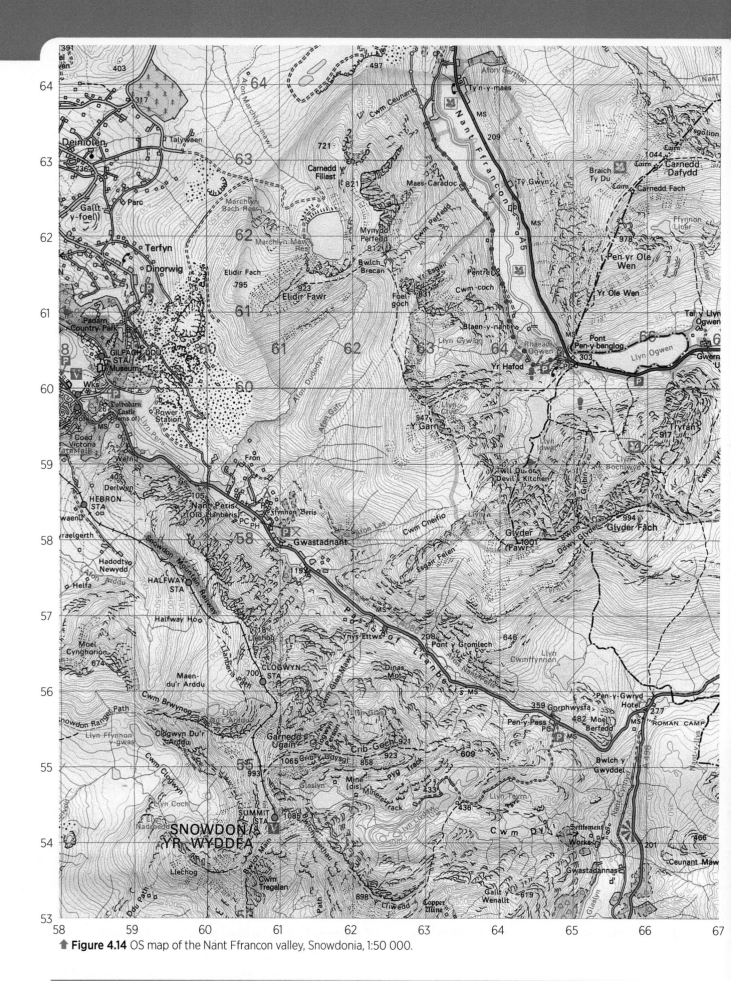

↑ Figure 4.14 OS map of the Nant Ffrancon valley, Snowdonia, 1:50 000.

Practise your skills

Use the 1:50 000 map of Snowdonia in Figure 4.14 on page 52 to help you answer these questions.

Figure 4.15 should also be used for question 1. The photograph was taken in grid square 6460.

1 In which direction was the camera pointing?
   north   south   east   west
2 Who owns the land on the valley floor?
3 The Nant Ffrancon valley is a glacial trough. Describe the valley. Remember to use data (figures) in your answer.
4 What type of woodland is found in grid square 5963?

5 How far is it, to the nearest kilometre, from Hebron station to Clogwin station along the Snowdon Mountain Railway?
6 Give a six-figure grid reference for each of the following glacial features:
   a) corrie
   b) arête
   c) truncated spur
7 Draw a cross-section of the Pass of Llanberis valley from 600555 to spot height 923 at 613613. Mark the following features on your cross-section: the A4086, Afon Nant Peris, the Snowdon Mountain Railway, Llanberis path and the National Park boundary.

⬆ **Figure 4.15** View of the Nant Ffrancon valley, Snowdonia.

What are the advantages and disadvantages of development in upland glaciated areas?

Upland glaciated areas have been developed in a number of ways. They have been used for water storage and supply, recreation and tourism and, more recently, for the development of **renewable energy**. The development of glaciated areas can have advantages and disadvantages both for the environment and for the people who live in the area. Many upland glaciated areas are also being designated as conservation areas, which can bring its own problems.

Recreation and tourism

Glaciated areas are used for recreation and tourism. This can have both advantages and disadvantages for the local area. It can also lead to changes to the glaciated landscape. A good example of this is Snowdonia National Park in Wales (see Figure 4.16).

Tourism supports rural services such as buses, village shops and post offices. Tourists use the services, which puts money into the local economy. In Snowdonia tourism is worth £396 million a year.

Local goods can become expensive because tourists will pay more, which raises the prices for local people.

Tourists come to see scenery and wildlife, which this puts pressure on areas to conserve them to ensure that tourists keep visiting.

The number of tourists visiting the area causes traffic congestion and problems for locals going about their daily business.

Snowdonia has nearly 2,500 footpaths. The number of tourists visiting the area causes footpath erosion, especially on the paths leading up to Snowdon.

Jobs in tourism are mainly seasonal, low paid and have long hours.

14% of Snowdonia's houses are holiday homes. The demand for holiday homes makes housing too expensive for local people.

8,300 people are employed both directly and indirectly by tourism in Snowdonia.

Tourists cause changes to the glaciated landscape in a number of different ways. They can cause erosion to footpaths because of the sheer number of people using them. This causes changes to the glaciated area as proper paths are put in due to the pressure of tourism.

Facilities are built for tourists, such as visitor centres, which have a visual impact on the landscape.

Snowdon has a railway to the summit, which has a major impact on the scenery of the glaciated area.

↑ **Figure 4.16** Snowdon's railway and footpath to the summit.

Conservation

Areas such as Snowdonia, which are popular with tourists, are usually conserved to protect their natural beauty (see Figure 4.17). This in itself can have advantages and disadvantages for the glaciated landscape and the people who live there. It can also result in changes taking place that are not always in keeping with the original glaciated landscape. In Snowdonia, the Conservation Snowdonia Project has been operating for four years. Its main aims are to stop the spread of non-native species, to maintain and repair footpaths on popular routes, and to co-ordinate voluntary conservation efforts. There are also a number of planning laws which are there to conserve the environment but can also have advantages and disadvantages.

Renewable energy

Glaciated areas are used for renewable energy, which can have both advantages and disadvantages for the local area (see Figure 4.8). It can also lead to changes to the glaciated landscape. An example of this is the wind farm at Whitelee, near Glasgow. This is the largest **onshore wind farm** in the UK with 215 wind turbines. It took ten years to build and was completed in 2009.

The removal of rhododendrons, which will take over an area, means that the glaciated landscape does not change, which is a positive outcome of conservation.

Maintaining footpaths in the Snowdon area can have disadvantages because, to some people, the footpaths can be a scar on the landscape. To others it means that most of the mountain is left natural and the visitors are kept to certain areas.

Special planning laws stop the demolition of properties that are not listed if they are within the national park. This is an advantage as it stops change, but it makes life complicated for the people who live in the park.

There are also restrictions on advertising, which is only allowed in certain areas.

An advantage of conservation is that it manages the development that takes place and ensures that it is done in a way that is in keeping with the local area.

Trees in the national park are also protected; any works on them needs permission. Again, this is a nuisance for people who live in the park.

Conservation can cause change to glaciated areas; for example, when footpaths are improved they can become a scar on the landscape.

Conservation sometimes means that settlements are not allowed to develop and are kept like a 'chocolate box' image, not the real world.

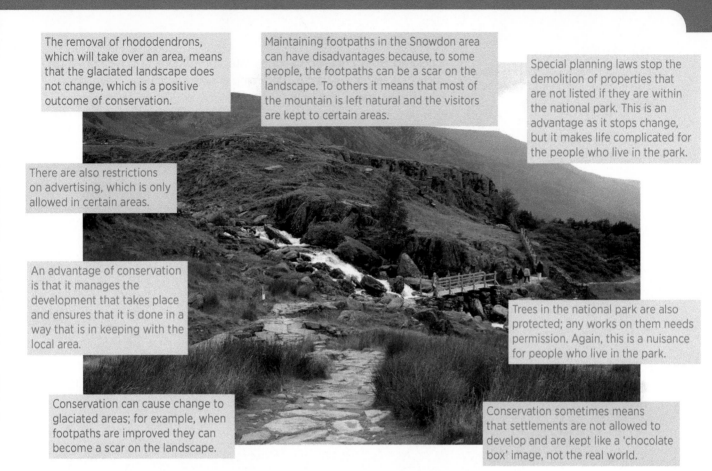

⬆ **Figure 4.17** A conserved footpath in Snowdonia.

There is a fear that some fauna will be disturbed by the turbines; for example, the noise they make can disorientate bats because the vibration interferes with their sonar.

The wind farm can generate 539 megawatts of electricity, which is enough to power about 300,000 homes.

There is a large visitor centre with an education centre, exhibition room, shop and a cafe with a viewing deck; there are cycling paths and well-marked walks. The views are amazing: 80 miles can be seen within a walking distance of 1.5 miles along gravel paths from the visitor centre.

Birds can be killed because they fly into the blades of the turbines.

Many people do not like wind farms because they are visually intrusive and ruin the natural look of the moorland; it can look industrialised with turbines, roads, control buildings and pylons.

Flora and fauna, including sphagnum moss and common lizards, will be protected on the site in a 2 500–hectare area of habitat management.

Wind turbines are quiet and aesthetically pleasing to many people.

There has been a loss of access to thousands of hectares of open moorland as there are now security fences and marked foot and cycle paths only in certain areas.

One of the main disadvantages is that the wind is unreliable.

The value of property in the area has decreased because people do not want to live in an area with a view of a wind farm.

Has a visual impact on the glaciated landscape because of the introduction of wind turbines.

The turbines have an impact on the flora and fauna of the glaciated landscape.

Tourists can be put off because the area is no longer as picturesque.

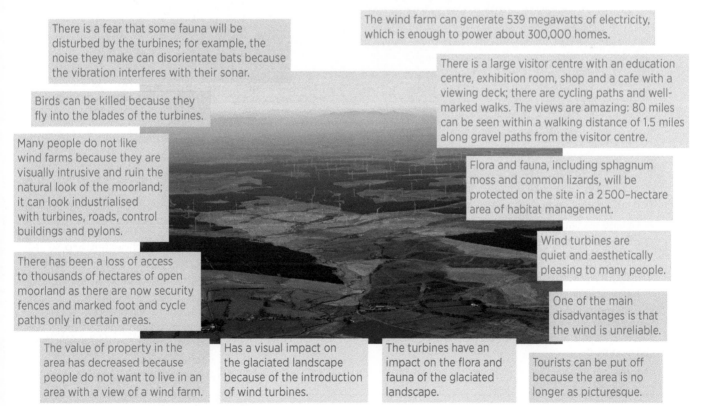

⬆ **Figure 4.18** Wind farm at Whitelee, near Glasgow.

SSSI – Site of Special Scientific Interest that is protected because of its unique habitat.

HEP – hydroelectric power.

⬇ **Figure 4.19** Kielder Water in Northumberland.

Water storage and supply

Glaciated areas are used to store water for human consumption. A large body of water is created by damming a valley, which can have both advantages and disadvantages for the local area. It can also lead to changes to the glaciated landscape. An example of this is Kielder Water in Northumberland, which was completed in 1982.

There are eight SSSIs in the Kielder area; it is one of last places for red squirrels in England.

£6 million is raised every year through tourism; it is visited by 250,000 tourists a year.

HEP is produced at Kielder Dam.

Kielder supplies water to settlements on the Tyne, Tees and Wear rivers, such as Newcastle and Middlesbrough.

The lake behind the dam can be used for recreational purposes, such as kayaking, windsurfing and sailing.

Kielder Forest employs 260 people; huge volumes of timber are also produced.

It flooded an area of scenic beauty.

It could be seen as visuall intrusive; the water and th forest are an enormous a of land, which could be seen as ugly and might p off some tourists.

200,000 million litres of water can be stored in the 11 sq km lake that was formed when the River North Tyne was dammed.

Kielder Village

Kielder Castle
• Visitor Centre

Kielder Forest has one sort of tree, Sitka spruce; 1.5 million trees were cut down to build the reservoir.

Visual impact on the glaciated landscape as the large body of water is not natural to that area.

Leaplish
Waterside Park •

River North Tyne

Minor Road

2,700 acres of farmlar and habitat were lost due to the lake.

• Tower Knowe
Visitor Centre

Glaciated landscapes are flooded with loss of land and settlements.

Calvert Trust •

It disrupts the natural processes at work in the glaciated landscape; sediment is trapped behind the dam causing lack of deposition further downstream.

Tourism can have an impact on the glaciated area, such as footpath erosion and traffic jams on crowded rural lanes.

58 families lost their homes under the water of the lake.

ACTIVITIES

1 State one way that farming has changed upland glaciated areas.
2 Upland glaciated areas are often used for water storage. Describe one advantage to the people of the area and one disadvantage to the environment of the area.
3 Study the information on how glaciated areas are being developed for renewable energy. Draw up a table like the one below and complete it with information from Figure 4.18 on page 55.

	Advantages	Disadvantages
Environment		
People		

Review

By the end of this section you should be able to:

✓ understand how farming, forestry and settlement have had an impact on upland glaciated landscapes
✓ know the advantages and disadvantages of the development of glaciated landscapes
✓ recognise that development can lead to change in glaciated landscapes.

Located example Distinctive glaciated upland landscapes are the outcome of the interaction between physical and human processes

LEARNING OBJECTIVE

To study how the interaction between physical and human processes produces distinctive glaciated upland landscapes.

Learning outcomes

▶ To know the significance of the location of the Isle of Arran.
▶ To know how physical processes formed the north of the island.
▶ To recognise how physical and human processes have changed the north of the island.

KEY TERMS

Highland boundary fault – a fault line across Scotland from west to east that divides two very different geological areas. It separates the highlands of Scotland from the lowlands of Scotland.

Granite batholith – a large area of igneous rock which forms below the Earth's surface.

Overgrazing – grass that is grazed so heavily that the vegetation is damaged and the ground becomes liable to erosion.

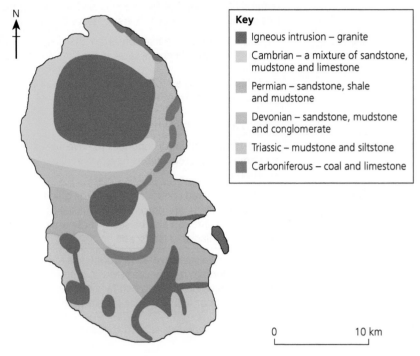

N

Key
■ Igneous intrusion – granite
  Cambrian – a mixture of sandstone, mudstone and limestone
  Permian – sandstone, shale and mudstone
  Devonian – sandstone, mudstone and conglomerate
  Triassic – mudstone and siltstone
  Carboniferous – coal and limestone

0   10 km

⬆ **Figure 4.20** A simplified geological map of the Isle of Arran, Scotland.

The Isle of Arran, which is located off the west coast of Scotland, has a distinctive glaciated landscape. The location is significant because the island is cut in two by the **highland boundary fault**, which is the line between the highlands and lowlands of Scotland. The island, therefore, is Scotland in miniature. The north of the island is a rugged mountainous area that is the result of an igneous intrusion. The south of the island has more gentle slopes with farmland and forests. The northern part of the island is a large **granite batholith** which was created about 60 million years ago. This large mass of molten magma pushed its way up into the Earth's crust, lifting the rocks above it by about 3,000 m. As these rocks were eroded by ice and water, the granite underneath was exposed. The area was then carved into its distinctive landscape by physical processes during the last ice age. The ice carved large glacial troughs, truncated spurs, corries, hanging valleys and arêtes using the processes of plucking and abrasion. It also deposited terminal and ground moraines. (For information on the formation of these glacial landforms see page 49). Since that time other physical processes, such as the work of rivers, have had an impact on the glaciated landforms producing the glaciated landscape that is there now.

Physical or human process	Impact on (change to) the upland glaciated landscape
Farming	The impact of sheep and, more recently, deer farming has changed the landscape. There are over 2,000 red deer in the north of the island. In the past the land was moorland but the increase in farming has meant that the pressure on the land has intensified. In some areas this has caused a loss of native habitats, which are sensitive to overgrazing.
Recreation and tourism	There are seven golf courses on Arran: the island is promoted as a golfing venue. This has changed the glaciated landscape as the links and fairways are not a native habitat for the island.
	Footpath erosion is a problem. Many of the local businesses charge extra for certain services in order to fund the Arran Trust, a charity that works to protect the environment.
	There is also the problem of litter and disturbance to livestock.
Freeze–thaw weathering	Freeze-thaw weathering continues to have an impact on the glaciated landscape. Scree slopes are common, as is rock debris from arêtes that are continually being sharpened by freeze-thaw weathering.
River processes	The rivers on Arran continue to erode the glaciated landscape. A number of river features, such as meander bends, are present in the glacial troughs (see Figures 4.2 and 4.22). Waterfalls and interlocking spurs can also be seen. These are changes that have occurred due to the work of rivers on the glaciated landscape.
Forestry	Originally Arran was covered in woodlands and moorlands. A quarter of the island is now covered in forests, mostly coniferous plantations (see Figure 4.23). This has had an impact on the landscape because the flora and fauna of these forests is different to the traditional deciduous forests and moorland that once covered the island.

⬆ **Figure 4.21** Physical and human processes that have shaped the landscape of the Isle of Arran.

⬆ **Figure 4.22** Physical processes at work on Arran, a meander bend.

⬆ **Figure 4.23** Changes to the upland glaciated landscape of Arran: coniferous tree plantation and deer fencing.

1 Identify two glacial landforms that can be seen in the north of Arran.
2 Describe how farming and forestry have changed the upland glaciated landscape.
3 Explain how physical processes have formed the upland glaciated landscape in the north of Arran.

Extension

Use the internet to find pictures of the upland glaciated landscape of Arran. Choose three and use them to explain how the work of glaciers and rivers has formed Arran's landscape.

The north part of Arran is formed from an igneous intrusion. Explain what is meant by an 'igneous intrusion'.

Review

By the end of this section you should be able to:

✓ recognise the significance of the location of the Isle of Arran
✓ know how physical processes formed the north of the island
✓ describe and explain how physical and human processes have changed the north of the island.

Practise your skills

Use the geology maps of the British Geological Survey (www.bgs.ac.uk/data/mapViewers/home.html) to complete your own geological map of the north of Arran.

On the Digimap website (http://digimap.edina.ac.uk) find past and present maps of Arran to investigate the changes to the north of the island, such as the amount of forestry.

Examination-style questions

1 Study Figure 4.8 on page 49. Identify **one** glacial landform shown in Figure 4.8.   (1 mark)
2 Study Figure 4.3 on page 45. State **one** type of weathering which might have an impact on this upland glaciated landscape.   (1 mark)
3 Water supply has both positive and negative effects on glaciated upland landscapes.
   Explain **one** way water supply has a positive effect on upland glaciated landscapes.   (2 marks)
4 Examine how physical processes work together to form the arête shown in Figure 4.9 on page 49.   (8 marks)

Total: 12 marks

Weather Hazards and Climate Chang

The atmosphere operates as a global system transferring heat and energy

What are the features of global atmospheric circulation?

The features of global atmospheric circulation are:

● The transfer of heat from the Equator to the poles.
● There are three **circulation cells** – Hadley, Ferrel and Polar.
● Jet streams impact on the movement of heat energy.
● The spin of the Earth creates the Coriolis effect.

KEY TERMS

Global atmospheric circulation – the worldwide movement of air which transports heat from tropical to polar latitudes.

Latitude – the distance north or south of the Equator. It is measured in degrees with the maximum being 90 °N or 90 °S.

Hemisphere – a half of the Earth. The northern hemisphere is above the Equator, the southern hemisphere is below the Equator.

Troposphere – the lowest layer of the atmosphere. It is thicker at the Equator (approximately 20 km) than at the poles (approximately 10 km).

Depression – a low-pressure system that produces clouds, wind and rain.

Ocean current – a continuous, directed movement of ocean water. The currents are made from forces acting on the water such as the wind, different temperatures and the Earth's rotation.

ITCZ – Inter Tropical Convergence Zone.

Trade winds – a wind that blows steadily from the tropics towards the Equator. In the northern hemisphere it is from the northeast and in the southern hemisphere from the southeast.

---

1 Hadley cells

Hadley cells stretch from the Equator to latitudes 30 °N and 30 °S. The features of Hadley cells are:

• Warm trade winds blow towards the Equator.

• At the Equator the trade winds from each hemisphere meet. The warm air rises rapidly causing thunderstorms. An area of low pressure is formed in the ITCZ where the air from the two cells meets over the Equator.

• The air at the top of the troposphere moves towards 30 °N and 30 °S where it becomes cooler and starts to sink back to the Earth's surface. As it descends, it warms and any moisture is evaporated. This creates high pressure areas, with cloudless skies. The world's hot deserts are found in these areas, such as the Sahara, in North Africa.

• On returning to the ground some of the air returns to the equatorial areas as trade winds; this completes the circle.

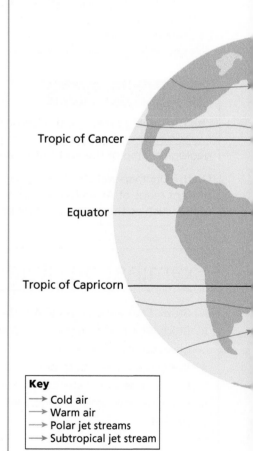

Tropic of Cancer

Equator

Tropic of Capricorn

Key
→ Cold air
→ Warm air
→ Polar jet streams
→ Subtropical jet stream

How do circulation cells and ocean currents transfer and redistribute heat energy across the Earth?

The main source of heat energy for the world is the Sun. The Sun heats the Earth's surface unevenly; it heats the Earth more at the Equator than at the poles. This creates a heat surplus at the Equator and a heat deficit at the poles. As the poles are not getting colder and the Equator is not getting noticeably warmer, there must be a redistribution of heat energy across the Earth. But how does this work? The heat energy is transferred in two ways: circulation cells and ocean currents.

Circulation cells

The three-cell model of global atmospheric circulation is shown in Figure 5.1. The air in each cell moves in a sort of circle. The cells transfer surplus heat energy from the Equator to the poles. In each hemisphere there are three pressure cells in which the air circulates through the troposphere: Hadley cells, which are closest to the Equator; Ferrel cells, and Polar cells, which are closest to the poles.

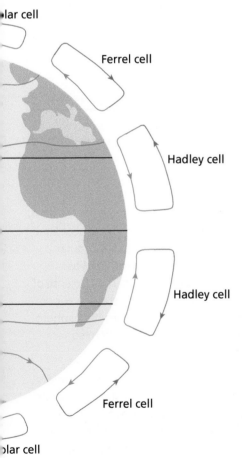

Polar cell

Ferrel cell

Hadley cell

Hadley cell

Ferrel cell

Polar cell

⬆ **Figure 5.1** The three-cell model of global atmospheric circulation.

② Ferrel cells

Ferrel cells stretch from latitudes 30 °N and 30 °S to latitudes 60 °N and 60 °S. The features of Ferrel cells are:

- Air on the surface is pulled towards the poles. This forms the warm southwesterly winds in the northern hemisphere and northwesterly winds in the southern hemisphere.
- These winds collect moisture as they blow over oceans on the Earth's surface.
- At about 60 °N and 60 °S they meet cold air from the poles.
- The warm air rises over the cold air as it is less dense. This produces low pressure at the Earth's surface and pressure systems known as depressions.
- Some of the air returns to the tropics and some is diverted to the poles as part of the Polar cells.
- The cell has a motion to the right in the northern hemisphere and to the left in the southern hemisphere due to the spin of the Earth. This is called the Coriolis effect.

③ Polar cells

Polar cells stretch from latitudes 60 °N and 60 °S to the north and south poles. The features of Polar cells are:

- The air sinks over the poles producing high pressure.
- The air then flows towards the low pressure in the mid-latitudes, about 60 °N and 60 °S. Here it meets the warm air of the Ferrel cells.

Jet streams

In the upper **atmosphere**, winds blow around the Earth in a westerly direction. Within these winds there are bands of extremely fast-moving air known as jet streams. These jet streams can be hundreds of kilometres in width, but only 1,000–2,000 m high. They are found at altitudes of about 10,000 m. The jet streams can be found in two areas of the world.

- Polar front jet stream – this is formed when cold Polar air meets warm tropical air high above the Atlantic Ocean, usually between latitudes 40° and 60°N and 40° and 60°S. Its exact location can vary. It marks the division between the Polar and Ferrel cells.
- The Subtropical jet stream – this is also generally in a westerly direction. It can be found at approximately 25°N and 35°S.

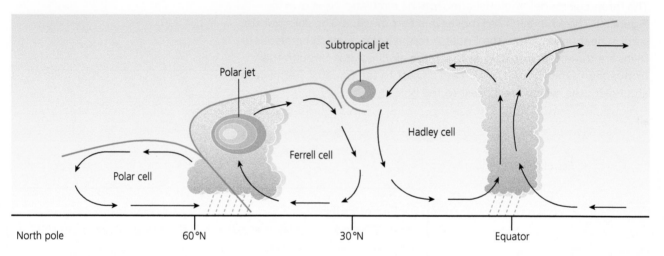

⬆ **Figure 5.2** A cross-section showing the location by latitude of the jet streams and the global circulation cells.

Ocean currents

The oceans transfer approximately twenty per cent of the total heat that is transferred from the tropics to the poles. Each ocean has a circular pattern of surface currents, known as a gyre. They are produced as masses of water move from one climatic zone to another. The currents of all the oceans are similar because they are created by the surface winds generated by global atmospheric circulation. In the northern hemisphere currents move in a clockwise direction and in the southern hemisphere they move in an anticlockwise direction.

The strongest currents are on the western side of oceans. For example, warm ocean currents such as the North Atlantic Drift transfers heat from low to high latitudes. This is particularly noticeable between latitudes 40° and 65° in winter, when warm winds blow onshore on the western sides of continents raising the temperature. Cold currents have less effect as they are usually offshore winds. One exception is the Labrador current off the east coast of North America.

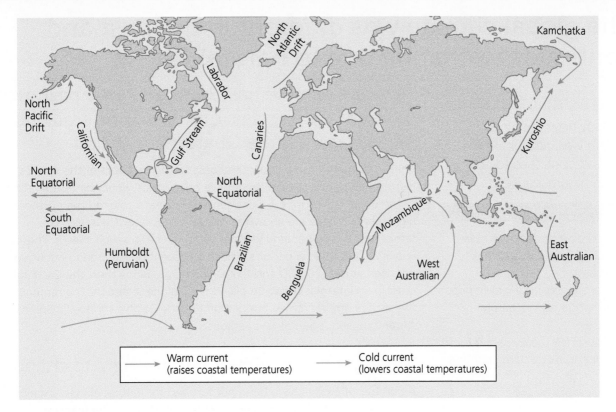

North Pacific Drift

Californian

North Equatorial

South Equatorial

Humboldt (Peruvian)

Gulf Stream

Labrador

North Atlantic Drift

Canaries

North Equatorial

Brazilian

Benguela

Mozambique

West Australian

Kamchatka

Kuroshio

East Australian

Warm current
(raises coastal temperatures)

Cold current
(lowers coastal temperatures)

⬆ **Figure 5.3** Major ocean currents.

ACTIVITIES

1. What is the main source of heat energy for the world?
2. Complete the following sentence.
   Ocean currents move in a .... direction in the southern hemisphere.
3. State three features of global atmospheric circulation.
4. Where are Ferrel cells found on the Earth's surface?
5. Describe how air moves in a Hadley cell.
6. Why are deserts found at approximately 25 °N and 25 °S of the Equator?

Extension

1. Explain how circulation cells and ocean currents redistribute heat energy across the Earth.
2. Research why low pressure is found at the Equator and high pressure is found at the poles.

Practise your skills

Study atlas maps to identify areas where the climate is affected by warm and cold ocean currents. Use the information to create a table.

Review

By the end of this section you should be able to:

✓ describe the main features of global atmospheric circulation

✓ explain how circulation cells and ocean currents transfer and redistribute heat energy across the Earth.

The global climate was different in the past and continues to change due to natural causes

■ LEARNING OBJECTIVE

To study how the global climate was different in the past and continues to change due to natural causes.

Learning outcomes

▶ To be able to describe how global climate was different in the past.

▶ To recognise how it continues to change due to natural causes.

▶ To describe the evidence there is for natural climate change.

How the climate has changed in the past over different time scales: glacial and interglacial periods during the Quaternary Period

The Quaternary Period covers the past 1.8 million years of the world's history. During this period there have been times when the temperature of the Earth has dropped. Continental ice sheets covered the northern hemisphere. These are known as glacial periods (or ice ages). The temperature then became warmer, melting the large ice sheets. These are known as interglacial periods. The most recent glacial period occurred between about 120,000 and 11,500 years ago. Since then, the Earth has been in an interglacial period called the Holocene epoch. The remnants of the last ice age still cover ten per cent of the Earth's surface in Greenland, Antarctica and mountainous regions.

What are the causes of natural climate change?

A number of causes have been put forward for natural climate change. These include **Milankovitch cycles**, **solar variation** and **volcanism**.

KEY TERMS

Climate – the average temperature and precipitation figures for an area.

Precession
The Earth's axis wobbles like a spinning top. A wobble cycle usually takes about 26,000 years. The motion is caused by the gravitational action of the Sun and the Moon. This cycle has an impact on the seasons and can cause warmer summers.

Precession
26,000 years

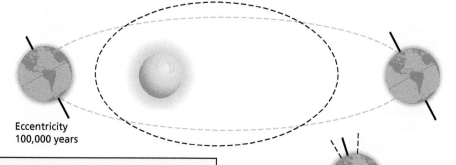

Eccentricity
100,000 years

Axial tilt
41,000 years

Eccentricity
The path of the Earth's orbit around the Sun is not a perfect circle, it is an ellipse. This shape can change from being nearly a perfect circle to more elliptical and then back again. This is due to the effect of the gravitational pull of other planets on the Earth such as Jupiter and Saturn. The measure of the shape's change is called its eccentricity. One complete cycle lasts for about 100,000 years. It appears that colder periods occur when the Earth's orbit is more circular and warmer periods when it is more elliptical.

Axial tilt
The Earth is spinning on its own axis. The axis is not upright; it tilts at an angle between 22.1°C and 24.5°C. A complete cycle for this tilt takes about 41,000 years. A greater degree of tilt is associated with the world having a higher average temperature.

⬆ **Figure 5.4** Milankovitch cycles.

Milankovitch cycles

These were developed by Milutin Milankovitch, a mathematician from Serbia. He put forward the theory that the amount of heat the Earth receives from the Sun is affected by its orbit. He identified three different cycles: eccentricity, axial tilt and precession, which are described in Figure 5.4.

Solar variation

This is a change in the amount of heat energy that comes from the Sun although these variations are very small and hard to detect. Sunspots on the Sun's surface do seem to have an impact on the heat energy of the Sun and therefore the climate of the Earth. There was a reduction in sunspot activity between 1645 and 1715 which corresponds with the Little Ice Age (see page 71). There has been a lot of sunspot activity since the 1940s, which could be a reason for the Earth's climate becoming warmer.

Volcanism

Large volcanic eruptions release ash and sulphur dioxide into the atmosphere. The ash quickly returns to Earth but the sulphur dioxide can have a cooling effect on the Earth's climate. The sulphur dioxide mixes with water in the atmosphere to form sulphuric acid droplets known as aerosols. These microscopic droplets absorb radiation from the Sun, heating themselves and the surrounding air. This stops heat reaching the Earth's surface. During the 1900s, there were three large eruptions that may have caused the planet to cool down by as much as 1°C. Eventually the effect will decrease as the aerosols fall as rain.

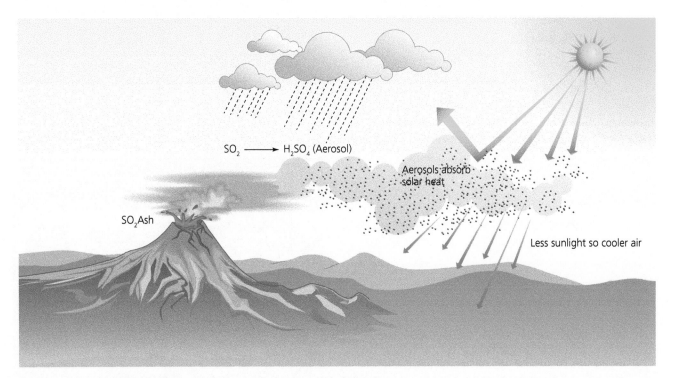

$SO_2 \longrightarrow H_2SO_4$ (Aerosol)

Aerosols absorb solar heat

SO_2 Ash

Less sunlight so cooler air

⬆ **Figure 5.5** How volcanic eruptions can have an impact on global climates.

What evidence is there for natural climate change?

Ice cores

The ice in areas such as Antarctica and Greenland has been there for millions of years. Cores can be drilled into it to measure the amount of carbon dioxide trapped in the ice. This is a climatic indicator because levels of carbon dioxide tend to be lower during cooler periods and higher when it is warmer.

Pollen records

Pollen analysis shows which plants were dominant at a particular time due to the climate. Each plant species has specific climatic requirements that influence their geographic distribution. Plants have a distinctive shape to their pollen grains. Pollen falls into areas such as peat bogs it resists decay. Changes in the pollen found in different levels of a bog indicate changes in climate over time.

Tree rings

Each year the growth of a tree is shown by a single ring. If the ring is narrow it indicates a cooler, drier year. If it is thicker it means the temperature was warmer and wetter. These patterns of growth are used to produce tree ring timescales, which give accurate climate information.

Historical sources

These include cave paintings, diaries and documentary evidence, for example, the fairs held on the River Thames when it froze. Since 1873 daily **weather** reports have been kept. Parish records are a good source of climate data for a particular area.

KEY TERMS

Weather – the day-to-day changes in temperature and precipitation.

Review

By the end of this section you should be able to:

✓ describe how climate has changed in the past over different time scales
✓ recognise the causes of natural climate change
✓ identify the evidence there is for natural climate change.

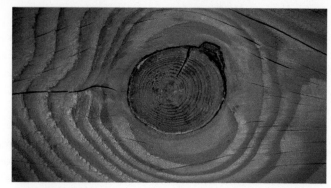
⬆ **Figure 5.6** Tree rings.

ACTIVITIES

1 What is meant by the term glacial period?
2 Describe three different forms of evidence for natural climate change.
3 How do volcanic eruptions cause the climate to change?
4 Describe two of Milankovitch's cycles.

Extension

Research how the climate has changed in the past. Produce a graph to show these climate changes.

⬆ **Figure 5.7** Illustration on the cover of a tract entitled *The Great Frost*. Cold doings in London, published in 1608.

Global climate is now changing as a result of human activity

How human activities produce greenhouse gases that cause the enhanced greenhouse effect

Human activities such as industry, energy, transport and farming produce carbon dioxide and methane. These gases are contributing to the enhanced greenhouse effect.

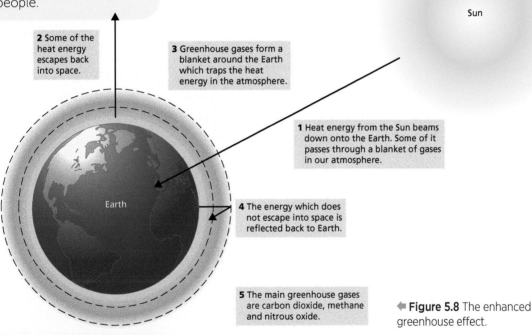

2 Some of the heat energy escapes back into space.

3 Greenhouse gases form a blanket around the Earth which traps the heat energy in the atmosphere.

Sun

1 Heat energy from the Sun beams down onto the Earth. Some of it passes through a blanket of gases in our atmosphere.

Earth

4 The energy which does not escape into space is reflected back to Earth.

5 The main greenhouse gases are carbon dioxide, methane and nitrous oxide.

⬅ **Figure 5.8** The enhanced greenhouse effect.

Sector	Carbon dioxide (m tonnes equivalent)	Methane (m tonnes equivalent)
Energy	180.8	7.6
Transport	115.7	0.1
Industry	12.2	0.1
Agriculture (farming)	4.9	27.0

⬆ **Figure 5.9** The contribution of energy, transport, industry and farming contribution to the enhanced greenhouse effect in the UK.

KEY TERMS

Enhanced greenhouse effect – also called climate change or global warming. It is the impact on the global climate of the increased amounts of carbon dioxide and other greenhouse gases that humans have released into the Earth's atmosphere since the Industrial Revolution.

Methane – fossil methane, which provides approximately 30 per cent of methane released into the atmosphere, was formed underground many years ago. It comes to the surface when fossil fuels are mined. Methane is a greenhouse gas; this means that it can trap heat within the Earth's atmosphere. It makes up twenty per cent of the greenhouse gases in the atmosphere and is twenty times more potent than carbon dioxide.

Industry

Some industrial processes contribute to the enhanced greenhouse effect, emitting large amounts of greenhouse gases such as carbon dioxide and methane. This occurs during the production process, for example during the production of iron and steel, chemicals and cement. In the UK the most prominent gas is carbon dioxide, and the most polluting process is the making of cement. Since 1990, however, there has been a 79 per cent reduction in greenhouse gas emissions from the industrial process sector in the UK. However, between 2012 and 2013 there was a rise in emissions as more cement, iron and steel were being produced.

Transport

Most forms of transport use **fossil fuels** to power them. When fossil fuels are burnt gases such as carbon dioxide are released, which build up in the atmosphere adding to the enhanced greenhouse effect. Since 1990 the emissions from transport in the UK have stabilised with a three per cent decrease overall. Passenger cars make up the largest part of this sector.

Energy

The generation of power accounts for 25 per cent of global carbon dioxide emissions. The main source is the use of coal and natural gas to produce electricity. For example, in China 75 per cent of energy is produced from coal. In the UK, since 1990 there has been a reduction in emissions of 32 per cent but the figure is still too high. The reduction has mainly been caused by a movement of power generation from coal as a fuel to gas.

Sector	USA greenhouse gas emissions (%)	UK greenhouse gas emissions (%)
Energy	31	33
Transport	27	21
Industry	21	12
Agriculture (farming)	9	9
Other (including residential)	12	25

⬆ **Figure 5.10** UK and USA percentage of greenhouse gas emissions per sector.

⬇ **Figure 5.11** Coal-fired power station.

Farming

Farming creates greenhouse gases in a number of ways, the main gas being methane. Livestock, especially cattle, produce methane as part of their digestion. This represents almost one-third of the emissions from the agriculture sector. Numbers of livestock in the UK are declining which has reduced emissions in this sector by nineteen per cent. However, in other parts of the world the numbers are increasing as there has been an increase in demand for Western-style diets which contain meat. An increase in rice production due to growing populations in Asia has also resulted in an increase in the production of methane. Manure is another producer of methane. New manure storage methods that allow less exposure to oxygen have reduced the amount of methane produced.

Sources of methane:
- wetlands including marshes and swamps
- the growing of rice
- landfills that contain rotting vegetable matter
- burning vegetation
- the bowels of animals
- mining of fossil fuels.

⬆ **Figure 5.12** Livestock can add to the greenhouse effect.

What are the negative effects that climate change is having on the environment and people?

Rising sea levels

Research published by the Intergovernmental Panel on Climate Change (IPCC) states that the global mean sea level has risen by between 10 and 20 cm over the last 100 years. According to The Met Office Hadley Centre for Climate Science and Services it has been rising by about 3 mm a year since the early 1990s. Predicting the amount of sea level rise in the future is more difficult. Recent studies predict a rise of between 0.8 and 2 metres by 2100. This would mean that a number of the world's largest cities would be under water such as New York. Other studies predict a meltdown of the Greenland ice sheet which would mean a sea level rise of 7 metres which would be enough to submerge London. In the worst case scenarios for sea level rise, between 665,000 and 1.7 million people who live on the Pacific islands of the nations of Tuvalu and Kiribati would have to find new homes.

⬆ **Figure 5.13** Rising sea levels.

Changing patterns of crop yield

Countries closest to the Equator are likely to suffer the most as their crop yields will decrease. In Africa, countries such as Tanzania and Mozambique will have longer periods of **drought** and shorter growing seasons. They could lose almost a third of their maize crop. It is forecast that in India there will be a 50 per cent decrease in the amount of land available to grow wheat due to hotter and drier weather.

Retreating glaciers

The vast majority of the world's glaciers are retreating (that is, melting), some more quickly than others. This is thought by some to be due to the increase in temperatures caused by climate change. Research has shown that 90 per cent of the glaciers in Antarctica are retreating. For example, between 1991 and 2013 the Sheldon Glacier, Adelaide Island, Antartica, retreated by 2 kms on its northern edge. The melting of the glaciers at the poles could also affect ocean water movement. It is believed that melting ice in the Arctic could cause the Gulf Stream to be diverted further south. This will lead to colder temperatures in western Europe, matching the temperatures found across the Atlantic in Labrador at the present time. Temperatures are frequently below 0 °C in the winter with averages of 8–10 °C in July, which is 10 °C cooler than the average UK summer temperature. The Columbia Glacier shown on satellite images in Figure 5.14 is in southern Alaska. It is one of the fastest retreating glaciers in the world. Between 1982 and 2014 it has retreated by 16 km and lost half of its thickness and volume.

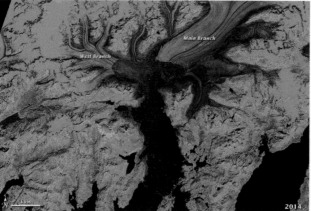

⬆ **Figure 5.14**
The retreat of the Colombia Glacier, Alaska.

Practise your skills

1 Draw a bar graph to show the data in Figure 5.10.
2 Describe the UK and USA's contribution to greenhouse gas emissions. Use data in your answer.
3 Check out the satellite images at http://earthobservatory.nasa.gov/Features/WorldOfChange/columbia_glacier.php to see how the Columbia Glacier is retreating. Think of the following: What impact will this have on sea level rises? What do you think is causing this retreat?

ACTIVITIES

1 Name two greenhouse gases.
2 Which sector in the UK produces the most carbon dioxide?
3 How do animals contribute to climate change?
4 Explain two negative effects of climate change.
5 How will changes to crop yields have an impact on food globally?

Extension

Examine how human activities can cause climate to change.

Review

By the end of this section you should be able to:

✓ describe how human activities produce greenhouse gases
✓ recognise the negative effects that climate change is having on the environment and people.

The UK has a distinct climate which has changed over time

LEARNING OBJECTIVE

To study how the UK has a distinct climate which has changed over time.

Learning outcomes

▶ To describe the climate of the UK today and to know how it has changed over the past 1,000 years.

▶ To understand the spatial variations in temperature, prevailing wind and rainfall within the UK.

▶ To recognise the significance of the UK's geographic location in relation to its climate.

How has the UK climate changed over the past 1,000 years?

The climate that the UK has experienced over the past 1,000 years has changed quite considerably. Between AD 800 and 1300, known as the medieval period, the UK experienced a warm period with temperatures around 1°C warmer than the average, see Figure 5.15. This is known because there are records of vineyards in Yorkshire, and agricultural productivity for the whole of the UK was high. Between 1300 and 1900 AD there was a period known as the 'Little Ice Age'. Temperatures were about 1°C cooler than present for most of the UK. The River Thames froze twenty times between 1564 and 1814 and ice fairs were held. In Scotland temperatures were 2°C cooler with prolonged winters and much snow. Between 1690 and 1700 the harvest in Scotland failed seven times, causing widespread famine. This is known because of parish records.

KEY TERMS

Precipitation – any form of moisture that reaches the earth; rain, snow, etc.

Maritime – influenced by the sea.

Annual temperature range – the difference between the highest and lowest temperatures of a place in a year.

Total annual rainfall – the sum of all the rainfall that falls in a year in an area.

Ice fairs – amusements held on the River Thames during the Little Ice Age.

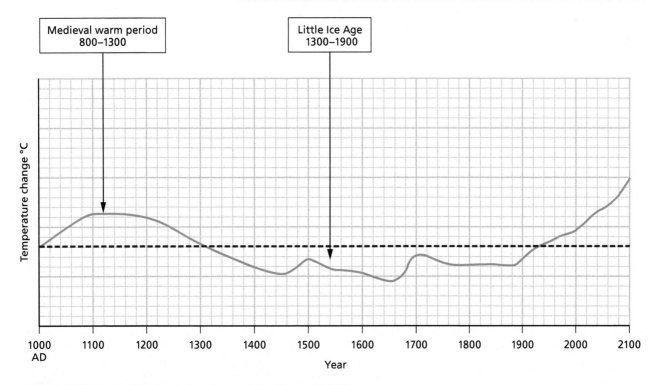

⬆ **Figure 5.15** How the UK climate has changed over the past 1,000 years.

What is the UK climate like today?

The UK climate today shows the characteristics of a **maritime** climate. The temperature does not have extremes and the **annual temperature range** is small, with maximum average temperatures of 15°C and minimum temperatures of 4°C. The temperature changes gradually between the months. The summers are warm and the winters are cool rather than cold.

Precipitation (mainly rain) falls every month; the total amount varies with location within the UK, from approximately 550mm in London to 1,800mm in Fort William. There is, however, little difference between the wettest and driest months. Figure 5.16 shows climate figures for different settlements in the UK.

KEY TERMS

Prevailing wind – the direction from which the wind usually blows. In the UK it is the southwest.

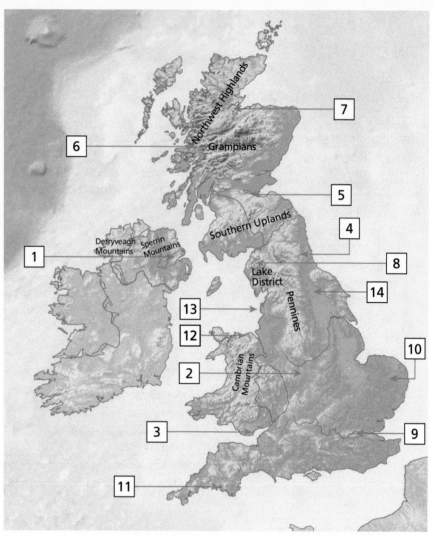

↑ **Figure 5.16** Climate data for UK settlements.

How does temperature, prevailing wind and rainfall vary within the UK?

As shown on Figure 5.16, there are variations in the temperature and rainfall of the UK depending on where you are located in the country. Settlements to the east of the country tend to receive less rainfall than settlements to the west. The main reasons for this are the type of rainfall the UK experiences and the relief of the country (see Figure 5.17). Temperature variations are the result of the influence of latitude and the distance the settlement is from the sea, as shown in Figures 5.19 and 5.20 on page 74. The **prevailing wind** of the UK is from the southwest, with little variation across the country. As this wind is blowing over the Atlantic Ocean it will bring rainfall to the UK. The predominance of this wind can be seen on the wind roses for different locations around the country shown on Figure 5.18 on page 5.73.

Number on map	Settlement	Max. temp (°C)	Min. temp (°C)	Rainfall (mm)
1	Belfast	13	6	950
2	Birmingham	13	6	800
3	Cardiff	15	7	1150
4	Durham	13	5	650
5	Edinburgh	13	6	700
6	Fort William	11	4	1800
7	Inverness	12	6	750
8	Keswick	13	6	1500
9	London	15	8	550
10	Norwich	14	6	650
11	Plymouth	14	8	1000
12	Rhyl	13	7	800
13	Blackpool	13	6	900
14	York	13	5	600

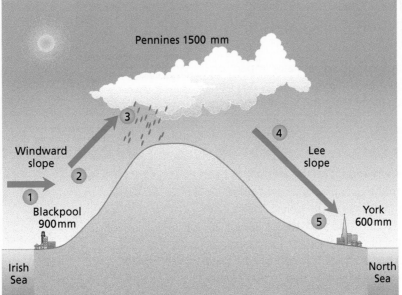

③ The water vapour in the air condenses to form clouds and it rains.

② Air rises up to pass over the hills. As it rises it cools by 1 °C every 100 m. Cool air cannot hold as much water as warmer air.

① Warm, wet winds blow towards the UK. They cross the Atlantic Ocean picking up moisture.

④ As the air moves down the hill it becomes warmer and can hold more water.

⑤ The wind blowing over York is gaining moisture, not losing it as rain.

Pennines 1500 mm

Windward slope

Lee slope

Blackpool 900mm

York 600mm

Irish Sea

North Sea

⬆ **Figure 5.17** An explanation of relief rainfall.

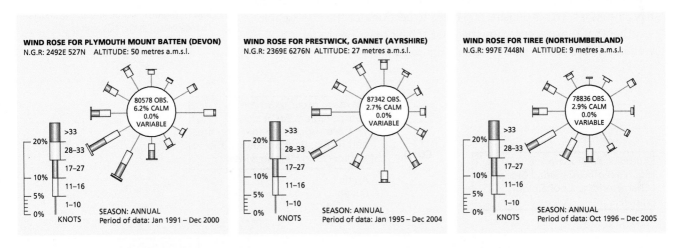

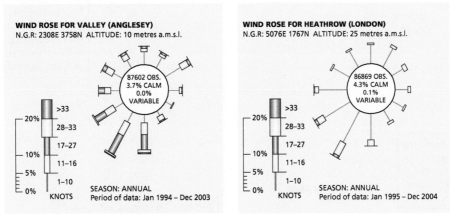

WIND ROSE FOR PLYMOUTH MOUNT BATTEN (DEVON)
N.G.R: 2492E 527N   ALTITUDE: 50 metres a.m.s.l.

80578 OBS.
6.2% CALM
0.0%
VARIABLE

>33
28–33
17–27
11–16
1–10
KNOTS

SEASON: ANNUAL
Period of data: Jan 1991 – Dec 2000

WIND ROSE FOR PRESTWICK, GANNET (AYRSHIRE)
N.G.R: 2369E 6276N ALTITUDE: 27 metres a.m.s.l.

87342 OBS.
2.7% CALM
0.0%
VARIABLE

>33
28–33
17–27
11–16
1–10
KNOTS

SEASON: ANNUAL
Period of data: Jan 1995 – Dec 2004

WIND ROSE FOR TIREE (NORTHUMBERLAND)
N.G.R: 997E 7448N   ALTITUDE: 9 metres a.m.s.l.

78836 OBS.
2.9% CALM
0.0%
VARIABLE

>33
28–33
17–27
11–16
1–10
KNOTS

SEASON: ANNUAL
Period of data: Oct 1996 – Dec 2005

WIND ROSE FOR VALLEY (ANGLESEY)
N.G.R: 2308E 3758N ALTITUDE: 10 metres a.m.s.l.

87602 OBS.
3.7% CALM
0.0%
VARIABLE

>33
28–33
17–27
11–16
1–10
KNOTS

SEASON: ANNUAL
Period of data: Jan 1994 – Dec 2003

WIND ROSE FOR HEATHROW (LONDON)
N.G.R: 5076E 1767N ALTITUDE: 25 metres a.m.s.l.

86869 OBS.
4.3% CALM
0.1%
VARIABLE

>33
28–33
17–27
11–16
1–10
KNOTS

SEASON: ANNUAL
Period of data: Jan 1995 – Dec 2004

⬆ **Figure 5.18** Wind roses for various areas within the UK.

What is the significance of the UK's geographic location in relation to its climate?

The UK's geographical location has a major impact on its climate due to a number of factors. These include latitude, air masses, distance from the sea and ocean currents.

Latitude

The latitude of the UK will impact on the amount of heat energy it receives from the Sun; places closer to the Equator are warmer than those at the poles. This is explained in Figure 5.19. Latitude also affects the temperature by influencing the length of the days. In the winter, the day length is short. This means that there are fewer hours of sunlight, resulting in lower temperatures.

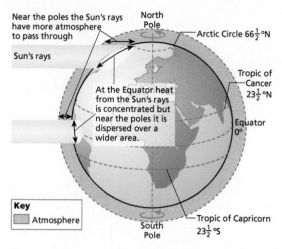

↑ **Figure 5.19** How latitude affects temperature.

Air masses

The geographical location of the UK means that its climate is influenced by five air masses. This is unusual and helps to account for the changeable weather that is experienced by the UK. Each air mass has different weather characteristics, as shown in Figure 5.20.

KEY TERMS

Source region – a large area of the Earth's surface where the air has a uniform temperature and humidity.

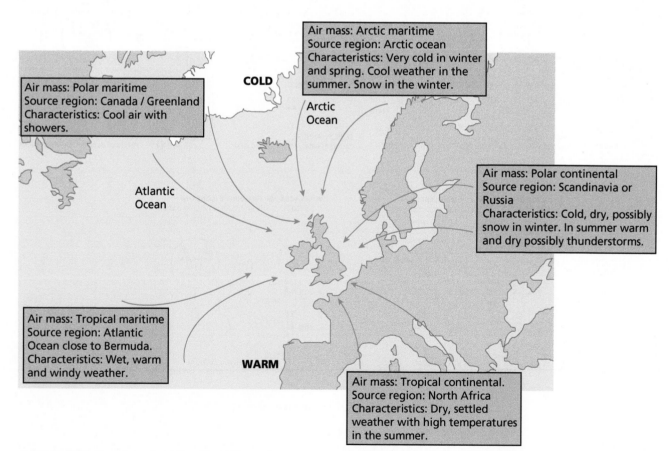

↑ **Figure 5.20** Air masses that affect the UK's weather.

Distance from the sea

The distance a settlement is from the sea has an effect on its climate. Settlements that are close to the sea will have less extreme temperatures than places further inland. In the winter, settlements close to the sea will be warmer than settlements inland, and in summer they will be cooler. The reason for this is that the land and sea respond differently to the heat energy from the Sun.

- The sea is constantly moving. Water that is heated or cooled on the surface circulates to a great depth. Therefore it takes a long time for the sea to heat up and cool down.
- The land is still. It is only heated to a depth of approximately 30 cm. Therefore, the land heats up and cools down quickly.

Ocean currents

Ocean currents also have an impact on temperatures. The UK is influenced by a warm current called the North Atlantic Drift. This current originates close to the Equator, moves through the Caribbean and across the Atlantic, and almost circles the UK. It is a warm body of water. It raises the temperature of the UK considerably as can be seen in Figure 5.21. The UK is further north than Boston and Montreal but its temperatures are much warmer in the winter due to the effect of the North Atlantic Drift.

City	Latitude	Av min temp (°C)
London	51	6
Boston	42	−2
Montreal	45	−9

⬆ **Figure 5.21** A comparison of lowest average temperatures.

Month	Temp (°C)	Rainfall (mm)
Jan	5	55
Feb	5	40
Mar	8	41
Apr	10	43
May	13	49
Jun	16	45
Jul	18	44
Aug	18	49
Sep	15	49
Oct	12	68
Nov	8	59
Dec	5	55

⬆ **Figure 5.22** Climate data for London.

ACTIVITIES

1 How could ice fairs be held on the Thames during the 1700s?
2 Calculate the mode and mean for London's rainfall.
3 Draw a dispersion diagram for London's rainfall using Figure 5.22. Mark on the diagram the median, the upper quartile, the lower quartile and the interquartile range.
4 State two reasons why Birmingham's temperatures are lower than Plymouth's temperatures.
5 Explain the importance of the UK's geographical position in relation to its climate.
6 Compare the information on the wind rose for Heathrow with the wind rose for Plymouth.

Extension

London is warmer in winter than Boston in the USA. Discuss this statement.

Practise your skills

- Use the data in Figure 5.22 to produce a **climate graph** for London.
- Describe the pattern of rainfall for London shown on your graph.
- Use an atlas to find climate graphs for different cities in the UK. Choose three and compare the temperature and rainfall information. Give a reason for each of your comparisons.

KEY TERMS

Climate graphs – graphs which show temperature as a line at the top and rainfall in bars beneath on the same graph.

Review

By the end of this section you should be able to:

✓ describe the climate of the UK today and to know how it has changed over the past 1,000 years
✓ understand the spatial variations in temperature, prevailing wind and rainfall within the UK
✓ recognise the significance of the UK's geographic location in relation to its climate.

Tropical cyclones are extreme weather events that develop under specific conditions and in certain locations

To study how tropical cyclones are extreme weather events that develop under specific conditions and in certain locations.

Learning outcomes

▶ To understand how global circulation in the atmosphere leads to tropical cyclones in source areas and the sequence of their formation.

▶ To know the characteristics, frequency and geographical distribution of tropical cyclones and how they change over time.

KEY TERMS

Non-frontal system – weather systems that do not contain a warm or a cold front; that are usually associated with mid-latitude low-pressure weather systems which bring rain.

What are the characteristics, frequency and geographical distribution of tropical cyclones and how do they change over time?

A tropical **cyclone** is a non-frontal intense low pressure weather system that is known by different names around the world , for example hurricane and typhoon, but have the same characteristics. Characteristics of tropical cyclones are:

- They develop over tropical and subtropical oceans between the Tropic of Cancer and the Tropic of Capricorn.
- They need a water temperature of over 27 °C to be able to form.
- They usually form towards the end of the summer and in the autumn.
- The highest number of storms occur in the North Pacific Ocean.
- The pressure gradient is very steep with close isobars.
- They feature strong winds and heavy rain, and often thunderstorms.
- The average wind speed is 120 kph but winds of 400 kph have been known.
- They normally move from east to west with the trade winds.
- They have an 'eye', which is the calm centre of the storm.
- Once they reach a wind speed of 60 kph they are given a name, each letter of the alphabet being used in turn. The lists rotate every six years. If a storm is particularly destructive the name is not used again and a new one is chosen. Floyd was replaced with Franklin in 2005.

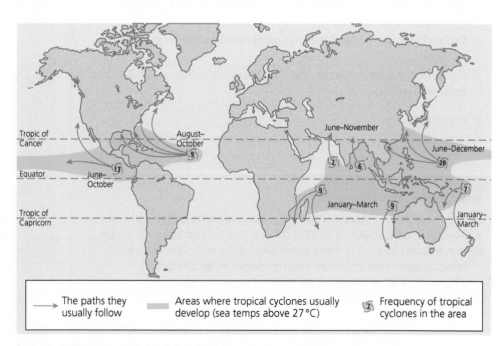

↑ **Figure 5.23** The geographical distribution and frequency of tropical cyclones.

How does global circulation in the atmosphere cause tropical cyclones and what is the sequence of their formation?

Tropical cyclones all start over oceans with a minimum temperature of over 27°C. Hot air rises taking a lot of water vapour with it. As it rises it cools to form big cumulus clouds. This creates low pressure at sea level. Air with higher pressure then moves in to replace it. This air does not move straight into the low pressure area because of the Earth's circulation, it whirls into it (just like water going down the plughole in a bath). This air then moves upwards with more water vapour. This has two effects: storm clouds are pulled into a spin by the incoming wind, and the spinning storm is pulled outward leaving a low pressure funnel, the eye, in the centre. The cold air, which is under high pressure, sinks down into the centre where it is heated and pulled into the spinning circle of air. The spinning circle begins to drift sideways because of the trade winds. This huge bundle of energy depresses the sea level under it, so there is a ridge of sea water giving rise to storm surges both before and after the cyclone has passed.

KEY TERMS

Source region – where tropical cyclones start their development.

Practise your skills

Describe the pattern of tropical cyclones shown on Figure 5.24.

⬇ **Figure 5.24** Stages in the formation of a hurricane.

1. Start of formation, western Caribbean, 22 October 2012.

Often causes thunder storms

Condenses into Cumulo-nimbus

Warm moist air rises

27°C Sea

Phew! It's hot in here

2. Formation of tropical depression, 23 October 2012.

Spins at 96 kph

I'm very sad, I'm a depression

300 km

3. Formation of Hurricane Sandy, 24 October 2012. Centre eye is calm.

Spins at 120–145 kph

My eye's in the wrong place!

400 km

4. Category 3 Hurricane Sandy arrives in Cuba, 25 October 2012.

Spins at 170–180 kph

Help, I'm going to be sick, I'm going too fast!

800 km

5. Category 3 Hurricane Sandy, effects on Cuba.

Trees and shrubs uprooted

Roofs blown off

People evacuated

Signs blown over

6. Category 1 Hurricane Sandy, leaves Cuba late on 25 October 2012.

Slowed back to 61–116 kph

Winds slower 120 kph

I'm dying!

ACTIVITIES

1 Copy out and complete the following sentences.
   a) Tropical cyclones need sea temperatures above .... to form.
   b) Tropical cyclones normally move from .... to west.
   c) Tropical cyclones normally form in the ....
2 State two other characteristics of tropical cyclones.
3 Draw your own annotated cartoon to show the stages in the formation of a hurricane.
4 Study Figure 5.23. In which months of the year does Australia experience tropical cyclones?

Extension

Research in depth what is meant by El Nino and El Nina.

Review

By the end of this section you should be able to:

✓ describe the characteristics, frequency and geographical distribution of tropical cyclones and how they change over time
✓ understand how global circulation in the atmosphere leads to tropical cyclones in source areas and the sequence of their formation.

There are various impacts of and responses to tropical cyclones depending on a country's level of development

Practise your skills

■ Use GIS and satellite images to track the path of Hurricane Sandy.
■ Use storm surge data to calculate the Saffir–Simpson magnitude of Hurricane Sandy.

Why are tropical cyclones hazards?

Tropical cyclones are hazards because of the high winds and intense rainfall that is experienced. The high winds can rip off house roofs and, in some cases, cause houses to simply fall over. During Hurricane Sandy in October 2012, up to 250 mm of rain fell in many places within just a few hours. This amount of rainfall will destroy crops and cause rivers to burst their banks. It can also cause landslides in many countries, which have been known to destroy whole villages.

Storm surges are caused when the high winds whip up the ocean's waves. Some waves experienced in Cuba during Hurricane Sandy were up to 10 m high and caused substantial coastal flooding.

Located example What were the social, economic and environmental impacts of Hurricane Sandy on Cuba and the USA?

On 23 October 2012, the government of Cuba warned the eastern states of the country about the imminent approach of Hurricane Sandy. The hurricane continued north affecting 24 states of the USA, causing particularly severe damage in New Jersey and New York (see Figure 5.27). The hurricane had different social, economic and environmental impacts on Cuba and the USA, however, because of the stage of development of the two countries.

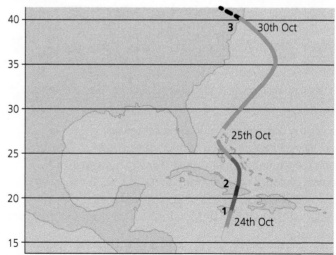

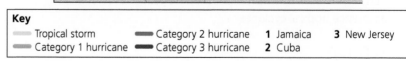

Key

▬ Tropical storm	▬ Category 2 hurricane	**1** Jamaica **3** New Jersey
▬ Category 1 hurricane	▬ Category 3 hurricane	**2** Cuba

⬆ **Figure 5.25** The path of Hurricane Sandy, October 2012.

The social, economic and environmental impacts of Hurricane Sandy on Cuba

1 Social impacts
- There was no electricity or fresh water.
- Eleven people were killed.
- Around 17,000 homes were destroyed and 226,000 were damaged.
- More than 55,000 people were evacuated because of the storm surge.

2 Economic impacts
- Total losses in the Santiago de Cuba area came to £50 million.
- Roads to the airport were blocked, so no tourists could arrive or leave the island, causing a loss in revenue.
- Total losses of US$2 billion.
- A five per cent drop in Cuba's GDP.

3 Environmental impacts
- Around 2,600 hectares of banana crops were destroyed.
- In Santiago de Cuba trees were uprooted and stripped of their leaves.
- Coffee plantations in mountainous areas were swept away.
- Areas close to the coast were flooded, with beaches being swept away, destroying wildlife habitats.

⬆ **Figure 5.26** The impact of Hurricane Sandy on the island of Cuba.

The social, economic and environmental impacts of Hurricane Sandy on the USA

① Social impacts
- 117 people were killed.
- Roughly nine million homes had power cuts.
- 650,000 homes were damaged or destroyed in the USA; 250,500 cars were destroyed by flood water.
- The streets of New York were flooded, as was the subway.

② Economic impacts
- Insurance claims in New Jersey totalled US$3.3 billion.
- US$1.1 billion was spent repairing the damage to sewage and water pipes in New Jersey and New York.
- The damage cost in New York totalled US$19 billion.

③ Environmental impacts
- The storm surge meant that sea water got into fresh water habitats, which had severe impacts on wildlife from Delaware Bay to Long Island Sound.
- Approximately 1.5 billion litres of sewage was released into the Raritan River in New Jersey.
- Around 90 per cent of beaches in New York and New Jersey were damaged; on average the beaches were 9–12 m narrower after the hurricane.
- 1.5 million litres of oil was spilt into Arthur Kill (the stretch of water between New Jersey and Staten Island, New York), damaging wildlife habitats and killing fish and birds.

⬆ **Figure 5.27** The impact of Hurricane Sandy in the USA.

KEY TERMS

CERF – the United Nations' Central Emergency Response Fund.

Humanitarian aid – help given after a natural disaster to save lives and reduce suffering.

FEMA – the USA's Federal Emergency Management Agency.

Located example How different were the responses of individuals, organisations and the governments of Cuba and the USA to Hurricane Sandy?

Responses by	Cuba	USA
Individuals	Many people moved in with relatives or friends; others took shelter in state workers' holiday homes where basic food was provided. They used materials provided by the government and other organisations to rebuild their own homes. The people of Cuba have no home insurance.	After the hurricane, people moved in with relatives and used shelters. People rebuilt their homes but used builders rather than doing it themselves. Most Americans have home insurance but those affected also received aid from the government and other organisations.
Organisations	The UN provided US$5.5 million to Cuba from the CERF and US$1.5 million in emergency funds. Venezuela sent 650 tonnes of aid, including non-perishable food, potable water and heavy machinery. Venezuela, Russia and Japan sent humanitarian aid. In the seven months following the hurricane, the Cuban Red Cross delivered support with the help of the Norwegian, Spanish and German Red Cross and Red Crescent Societies. The relief aid went to approximately 25,000 families and included roofing materials, mattresses, clean drinking water, and hygiene and kitchen kits. The World Food Programme responded immediately with US$1 million to assist the 788,000 people in the worst affected areas of Cuba with a one-month food ration from December 2012 to February 2013.	Hurricane Sandy caused extensive erosion to the Delaware Bay beaches, which had an impact on the breeding grounds of horseshoe crabs. The Canadian Rivers Institute worked with a number of other NGOs and public agencies to restore these beaches by clearing rubble and replenishing sand to provide a nesting area for horseshoe crabs. The Red Cross had 17,000 trained workers, 90 per cent of them volunteers, providing over 300 response vehicles, 74,000 overnight stays, and 17 million meals and snacks, among other aid. AmeriCares, an American charity, responded quickly by sending teams of relief workers to hard-hit areas, sending aid shipments, providing funding and deploying a mobile medical clinic. In the two years following Hurricane Sandy, AmeriCares has provided US$7.1 million in aid benefiting 450,000 people.
Governments	The government sent teams of electricians from all over the island to Santiago province within hours of the hurricane hitting. The government announced a 50 per cent price cut for construction materials and interest-free loans to repair the damage. The aid will be means tested and more subsidies will be available for the poorest or hardest hit. The government made building materials available to residents, including corrugated iron sheets, metal rods and cement. Local government officials compiled data from families about the damage they had suffered so that the government could send the appropriate help. Military teams were mobilised quickly to clear the streets of rubble and an estimated 6.5 million m³ of felled trees.	The US government approved US$60.3 billion in aid to the victims of Hurricane Sandy. The government promised that there would be improved weather forecasting, especially of storm surges. FEMA teams and resources were put in place to help people even before the hurricane had caused any problems. They were on hand to offer any help that was needed. FEMA and the Army Corps of Engineers worked with state and local governments to quickly reopen most of the beaches in New Jersey. The Department of Agriculture promised US$6.2 million for emergency food assistance, infrastructure and economic programmes to help repair farmland and flood plains.

⬆ **Figure 5.28** The responses of individuals, organisations and the governments of Cuba and the USA to Hurricane Sandy.

Review

By the end of this section you should be able to:

✓ explain why tropical cyclones are hazards
✓ understand the different social, economic and environmental impacts that Hurricane Sandy had on the USA and Cuba
✓ recognise the different responses that individuals, organisations and the governments of the USA and Cuba had to Hurricane Sandy.

ACTIVITIES

1 State four characteristics of tropical cyclones.
2 Explain why tropical cyclones can be hazards.
3 Use Figure 5.25 to describe the path of Hurricane Sandy.
4 Construct a table showing two economic, environmental and social impacts of Hurricane Sandy on Cuba and the USA.
5 How did the help given by the government of Cuba differ to that given by the government of the USA?
6 Choose two of the responses by organisations to the disaster in Cuba. Explain why this help was given.

Extension

The impacts of Hurricane Sandy were different in Cuba than in the USA. Examine three reasons for these differences.

Practise your skills

Research the impact of Hurricane Sandy on Cuba and the USA. Ensure that you use each of the following types of information:

- socioeconomic data to assess the impacts on different countries
- aerial, oblique and ground photographs to interpret the damage caused by the hurricane
- social media, such as blogs, and news sources for comments on the impact of Hurricane Sandy
- satellite images to assess the impact of Hurricane Sandy.

The causes of drought are complex, with some locations more vulnerable than others

LEARNING OBJECTIVE

To study the causes of drought and to understand why some locations are more vulnerable than others.

Learning outcomes

▶ To describe the characteristics of arid environments compared to the characteristics of areas suffering from drought.

▶ To understand the different causes of drought: meteorological, hydrological and human.

▶ To explain why global circulation makes some locations more vulnerable to drought than others and how this changes over time.

What are the characteristics of arid environments compared with those suffering from drought?

The characteristics of arid environments are:

- an average rainfall of between 100 and 300 mm
- variations in rainfall totals of between 50 and 100 per cent each year
- **pastoral farming**, usually by **nomadic herdspeople**
- natural vegetation is sparse – grasses, small shrubs and trees
- a short growing season of about 75 days.

Areas suffering from **drought** can be in arid environments but can also be in other climate regions. Therefore, the main characteristic of an area suffering from drought is a gradual reduction in the amount of available water supply. By contrast, a characteristic of arid environments is the fact that there is *always* only a small amount of rainfall and, therefore, water available for human use. It is the gradual nature of droughts which makes them so dangerous. It is also the fact that its occurrence is unpredictable and, to some extent, unexpected. This is shown in more detail in the examples of the droughts in California and Namibia at the end of this chapter.

KEY TERMS

Pastoral farming – the rearing of sheep, cattle, pigs or any other animals on a farm.

Nomadic herdspeople – people raising animals for their own food; they move around and have no fixed land.

Drought – a period of below-average precipitation resulting in prolonged shortages in water supply.

⬆ **Figure 5.29a** An arid environment.

⬆ **Figure 5.29b** An area suffering from drought.

What are the different causes of drought?

Causes of drought	Definition
Meteorological	This concerns the amount of precipitation an area receives compared to its average. It is all about the weather and occurs if there is a prolonged period of below-average precipitation, which creates a natural shortage of available water; this is then called a drought.
Hydrological	This is how a decrease in precipitation can have an impact on overland flow, reservoirs, lakes and ground water. This is often defined on a river basin scale: water reserves in aquifers, lakes and reservoirs fall below an established statistical average This can be related to precipitation or human demand and increased usage.
Human – agricultural	This is when there is not enough water available to support crop production on farms. This can occur when the crops are planted or during their growing season; it often occurs when there is a fall in precipitation but can also occur if farming techniques change. For example, farmers could use irrigation to start growing crops that require more water than is available; if the irrigation source dries up then the plants will die.
Human – dam building	If a dam is constructed on a large river it can produce electricity and plenty of water for the area close to the dam. However, places further downstream may suffer from drought because they will be receiving a reduced flow of water. For example, the building of the Atatürk Dam on the River Euphrates provided electricity and water for irrigation in Turkey but has restricted the flow of water to Syria and Iraq, meaning they have less water for irrigation.
Human – deforestation	The cutting down of trees for fuel reduces the soil's ability to hold water. This can cause the land to dry out, which can result in drought in an area.

⬆ **Figure 5.30** Natural and human causes of drought.

Why does global circulation make some locations more vulnerable to drought than others?

The movement of air from the Equator to the middle latitudes is discussed on page 60. The air rises at the Equator causing thunderstorms and a loss of moisture. The drier air moves north towards the mid latitudes. When it reaches approximately 30°N and 30°S; the dry air descends and warms. Many of the world's arid areas are found at these latitudes because of this air circulation pattern. The climate regions aren't fixed, however, and studies have found that global winds are shifting: it is thought that the jet streams have moved towards the poles, and the tropical belt has widened by several degrees latitude since 1979. This means that the arid areas of the world are changing. For example, the Upper Colorado River Basin, which is suffering from drought, is at latitude 37°N.

Review

By the end of this section you should be able to:

✓ describe the characteristics of drought and arid environments
✓ explain the different causes of drought: meteorological, hydrological and agricultural
✓ explain why global circulation makes some locations more vulnerable to drought than others.

The impacts of and responses to drought vary depending on a country's level of development

Why are droughts hazardous?

There are a number of reasons why droughts are hazardous. People and environments cannot survive without water; if there are drought conditions, it is likely to become a natural disaster. Droughts are different to other natural hazards in that they develop slowly over many years. They can have devastating effects.

- There will be a shortage of water supplies and residents will be asked to conserve water. In the UK water supplies have been switched off for periods of the day and residents have had to use stand pipes in the street to obtain fresh drinking water. If these conditions persisted, it could lead to diseases associated with poor living conditions. In developing countries the lack of clean drinking water due to droughts can lead to deaths from diseases such as cholera.
- Crops fail during droughts and animals die due to the lack of grazing land. This lack of food causes malnutrition and ill health among the population of the area experiencing the drought.
- The environment can also be destroyed by drought. When soil becomes dry due to lack of rain, vegetation dies leaving the soil unprotected. The dry soil can then be blown away by the process of wind erosion. When rain returns to the area there is no top soil left, so natural vegetation cannot regrow and crops cannot be planted.
- Wildfires are more common during periods of drought. This is because the trees are very dry and burn easily. There will also be a lot of fallen branches and dead wood lying in the forest. If a fire does start, water supplies to help control it will also be scarce.

Located example Impacts and responses to a drought in a developing country: Namibia, drought of 2013

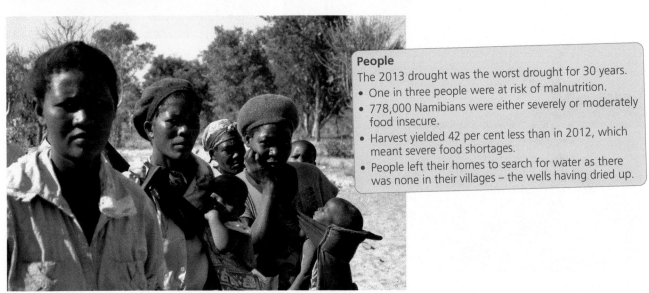

People

The 2013 drought was the worst drought for 30 years.
- One in three people were at risk of malnutrition.
- 778,000 Namibians were either severely or moderately food insecure.
- Harvest yielded 42 per cent less than in 2012, which meant severe food shortages.
- People left their homes to search for water as there was none in their villages – the wells having dried up.

Responses by organisations

- UNICEF appealed for US$7 million to support their efforts to respond to the needs of those affected.
- The International Red Cross and Red Crescent asked for US$1.5 million.
- Algeria donated US$1 million in food aid.
- The Lutheran Church helped in a number of ways. For example, by providing basic food assistance to vulnerable communities with no access to governmental distribution points, trying to ensure a safe environment and access to clean water.

Responses by individuals

- Farmers been forced to sell their livestock.
- People migrated to towns in search of work.
- In one village almost all the people, about 350, left in search of water and grazing land for their cattle.

Responses by the government

- In May, President Pohamba declared a state of emergency and requested US$1 million in international support to avert a crisis.
- Pledged £13 million in relief for the worst-hit households.
- The Ministry of Agriculture, Water and Forestry (MAWF) gave two options to farmers who do not have enough grazing for their animals: either to sell their livestock while they are still in good condition or a subsidy for the cost of transporting their animals to emergency grazing areas.

Environment

- Severe drought can have a great impact on a savannah ecosystem. It can change an area of grassland that could sustain livestock to an area of inedible grasses and plants that livestock cannot live on. This is because the grasses that can cope with drought are not good for livestock.
- Large areas of Namibia are changing from savannah grasslands to desert due to the lack of rainfall. Only drought-resistant plants can survive in these conditions but the Namibian farmers' cattle cannot graze on them.

⬆ **Figure 5.31** The impacts and responses to drought in Namibia.

Located example Impacts and responses to a drought in a developed country: California, drought of 2014

People

The effects on people can be seen in this photo, which indicates how people were restricted in their usage of water.

- Farmers use 80 per cent of the human usage of water. If there are water shortages their crops will die or they will plant fewer crops, resulting in less food for people.
- Loss of 17,100 jobs in farming.
- 5 per cent of the irrigated land in California won't be planted.
- The Department of Agriculture predicts that prices of fruit and vegetables will rise by 6 per cent.

Environment

A lack of water will have an impact on the environment of California in many ways.

- Wildfires are becoming a regular occurrence because of the dry and dead wood. There were more than a dozen fires in May 2014 near San Diego. Fires of this nature usually occur in the autumn.
- Some rivers and streams are closed to fishing. If water levels continue to drop the water will become warmer and the young salmon will be unable to survive as they require cool running water.
- The earth is shrinking because of depleted groundwater reserves.
- Between 2008 and 2011, parts of the Central Valley subsided more than 60 cm.
- The dry weather meant a better grape harvest and better tasting wines.

Responses by the government

- The state is preparing to undertake fish rescues – capturing them in shallow waters and transporting them to deeper waters closer to the ocean. For example, Chinook salmon have been taken in trucks to San Pablo Bay because there was little water for them to swim through to get to the ocean.
- Governor Brown issued a state of emergency.
- In February President Obama gave US$183 million from federal government funds.
- In March Governor Brown signed drought-relief legislation worth US$687 million. It included US$25.3 million for food and US$21 million for housing for people such as farmworkers who are out of work.
- Residents in the Santa Clara Valley have been told to limit watering of lawns to twice a week or face a US$500 fine.
- The Reclamation and Natural Resources Conservation Service announced that they hope to provide up to US$14 million of federal funds to help farmers to conserve water and improve water management.

Responses by organisations

- New mandatory laws forbid restaurants to put water on tables without it being requested.
- Hotels must also ask guest if they will reuse their linen and towels to save water.
- Advanced forecasting models are being developed by NASA to help with the water shortage.
- They are also developing new ways to better manage and monitor the state's water resources.

Responses by individuals

- Farmers will have to pump more water, which will cost an extra US$453.
- People have been asked to use water more sparingly.
- Farmers are planting smaller crops because there is not enough water for them to grow.

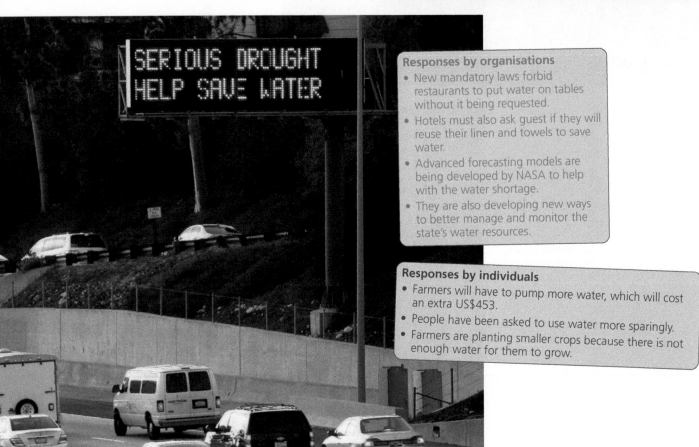

⬆ **Figure 5.32** The impacts and responses to drought in California.

ACTIVITIES

1 State one human and one physical cause of drought.
2 Draw a table that compares the impact of drought on people and the environment in the USA and Namibia.
3 State one impact of drought for people living in the USA.
4 Compare the response to drought by organisations in the USA and Namibia.

Extension

Evaluate the responses of individuals and governments to drought in developing and developed countries.

Review

By the end of this section you should be able to:

✓ explain the reasons why droughts are hazardous
✓ describe the impacts of drought on people and ecosystems in Namibia and the USA, and how they differ between the two countries
✓ compare and contrast the different responses to drought from individuals, organisations and governments in Namibia and the USA.

KEY TERMS

Irrigation – the artificial watering of land for farming.

Examination-style questions

1 a) State which of the following is a greenhouse gas.
   A oxygen   B nitrogen   C carbon monoxide D methane   (1 mark)
   b) State **two** human causes of climate change.   (2 marks)
   c) Explain **one** negative effect that climate change is having on people.   (3 marks)
2 Study Figure 5.3 on page 63.
   a) State if the California current is a warm or a cold current.   (1 mark)
   b) State which ocean current has an effect on the climate of the UK.
   A Canaries   B Gulf Stream   C Labrador   D North Atlantic Drift   (1 mark)
   c) Ocean currents influence the climate of the UK.
      Explain **one** way that ocean currents affect the climate of the UK.   (3 marks)
3 The UK's climate is very variable. This is shown on Figure 5.16 on page 72.
   a) Calculate London's annual temperature range (see Figure 5.22).   (1 mark)
   b) Calculate the median maximum temperature shown on Figure 5.16.   (1 mark)
   c) Blackpool has a higher total annual rainfall than York (see Figure 5.17 on page 73).
      Suggest reasons why?   (3 marks)
4 Study Figure 5.1 on pages 60 and 61.
   a) State **two** features of global atmospheric circulation.   (2 marks)
   b) Explain how air moves in the Polar circulation cell.   (4 marks)
5 Evaluate the different responses to the impact of a tropical cyclone on people in a named developing country.
   Named developing country ....   (8 marks)

Total: 30 marks

Ecosystems, Biodiversity and Management

▮ LEARNING OBJECTIVE

To study large-scale ecosystems.

Learning outcomes

▶ To know the distribution of the world's large-scale ecosystems.
▶ To be able to describe the characteristics of the world's large-scale ecosystems.
▶ To understand the role of climate and local factors in influencing the distribution of large-scale ecosystems.

Large-scale ecosystems are found in different parts of the world and are important

What is the distribution and characteristics of the world's large-scale ecosystems?

The world's large-scale **ecosystems** are tropical forests, **temperate forests**, **boreal forests**, tropical grasslands, temperate grasslands, **deserts** and **tundra**. The **distribution** of these ecosystems is shown in Figure 6.1, along with the characteristics (or main features) of each ecosystem and a country where it can be found.

⬇ **Figure 6.1** The distribution and characteristics of the world's large-scale ecosystems.

Tundra – Canada
• Temperature range between −34 °C and 12 °C
• Total annual rainfall 200 mm, much of which falls as snow
• Short growing season of about 60 days
• Permafrost – permanently frozen ground
• Very poor surface drainage
• Plant species have shallow root systems and are low to the ground to cope with the harsh climate, such as mosses, lichens, grasses and dwarf shrubs. They are low to the ground and have a small leaf structure so that they can repel the cold temperatures. Animals that live here have adapt in different ways – brown bears eat in the summer and then store the food in a thick layer of insulating fat which they live off while they are hibernating in the winter.

Boreal forests – Russia
• Temperature range between −10 °C and 15 °C
• Total annual rainfall 500 mm.
• Trees have a thick bark to protect them from the cold
• Needle leaves to slow down transpiration
• Evergreen trees, which allows growth to start when the weather warms up
• Shallow root systems because of shallow soil and frozen ground
• Trees such as pine and fir
• Animals such as red foxes and black bears.

Deserts – Australia
• Temperature range between 30 °C and 35 °C
• Great temperature differences between day and night, −18 °C and 45 °C
• Very unpredictable rainfall, but generally very low
• Sand or very coarse soils with good drainage
• The only plants are short shrubs such as the prickly pear cactus, which stores water in its spongy tissue
• Animals such as camels live in the desert, which store fat in their humps which they can change into water when it is needed.

Temperate forests ecosystem - USA
• Temperature range between 4 °C and 17 °C
• Total annual rainfall 1,000 mm
• Trees lose their leaves in winter to reduce transpiration
• Vegetation is in 4 layers – canopy, sub-canopy, herb and ground. Many are dominated by one tree species – Oak.
• Animals such as rabbits and deer.

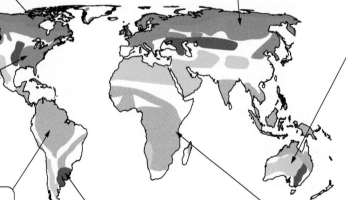

Key
▮ Boreal forest
▮ Temperate forest
▯ Tropical forest
▮ Tundra
▮ Temperate grasslands
▮ Tropical grasslands
▯ Hot deserts

Tropical forests – Brazil
• Temperature range between 27 °C and 30 °C
• Total annual rainfall 2,200 mm
• Soil: poor quality; nutrients washed down through the soil due to the amount of rainfall, forming a hard pan (a layer of solid nutrients lower down in the soil which cannot be accessed by plants)
• Vegetation is in four layers: emergents, canopy, under canopy, shrub/forest floor
• Lianas wind their way up the trees
• Epiphytes grow on the trees
• The forest is evergreen
• Animals such as sloth and howler monkeys.

Temperate grasslands – Argentina
• Temperature range between 10 °C and 18 °C
• Total annual rainfall 500 mm, most falling in the summer months
• Trees are generally not found in these areas
• Grasses such as purple needlegrass and buffalo grass grow in these areas
• The temperate grasslands in North America, known as the Prairies, have been converted into farmland.

Tropical grasslands – Kenya
• Temperature range between 25 °C and 30 °C
• Total annual rainfall 1,000 mm
• Rain is concentrated in 6–8 months of the year, the rest of the year has drought conditions
• Animals, for example giraffes, reproduce during the wet season when there is plentiful food and water
• Grasses grow very tall during the wet season, up to 2 m high, but die off during the dry season
• A few trees are found in these areas such as the acacia tree, which survives due to its thick trunk which holds water.

How does climate and other more local factors, such as soils and altitude, influence the distribution of large-scale ecosystems?

The distribution of large-scale ecosystems is determined to a great extent by climate. As shown in Figure 6.1, the different ecosystems have different climatic requirements. For example, tropical rainforests require warmth and moisture for plants to survive. Boreal forests are found in areas that have a short growing season and are made up of tree species that do not lose their leaves when the weather becomes colder. **Deciduous** forests would not survive in these areas because the trees would not have time to regrow their leaves.

Altitude is another factor that influences the type of vegetation because, as the land becomes higher, a number of changes take place in the **abiotic** factors.

- The temperature drops by 1°C per 100 m.
- **Soils** become thinner and contain less **organic** matter.
- The soil temperature also drops.

This has a number of impacts on the vegetation. Plant size decreases and there are more grasses and fewer trees. There is less diversity of both plants and animals species, and plants have a lower growth rate. The degree to which these trends happen depends on where the mountain is located. Figure 6.2 shows the possible changes for a mountain close to the Equator.

Soils are also important because different plants require different types of soil. The underlying bedrock and soil are therefore important for the type of ecosystem found in an area. For example, boreal forests have a very acidic soil because of the leaf **litter** from the trees. Deciduous forests have a more nutrient-rich litter because of the leaf fall every autumn.

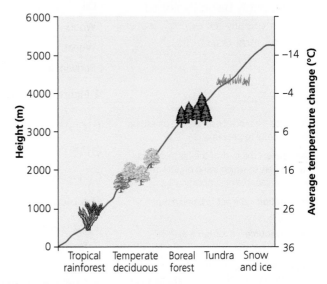

⬆ **Figure 6.2** Changes in vegetation type and temperature as altitude rises.

Review

By the end of this section you should be able to:

✓ recognise the distribution of the world's large-scale ecosystems
✓ describe the characteristics of the world's large-scale ecosystems
✓ explain the role of climate and local factors (soils and altitude) in influencing the distribution of large-scale ecosystems.

KEY TERMS

Litter – decomposing leaf and other organic debris found on forest floors.

Ecosystems – a community of plants and animals and their non-living environment.

Distribution – where something is located.

Altitude – height above sea level.

ACTIVITIES

1 Complete the following table for all of the ecosystems shown in Figure 6.1.

Ecosystem	Location	Main characteristics
Tropical forests		
Temperate forests		
Boreal forests		
Tropical grasslands		
Temperate grasslands		
Deserts		
Tundra		

2 Explain how altitude affects the distribution of ecosystems.

Extension

Altitude has an impact on ecosystems. Explain how the location of a mountain also has an impact on the ecosystems which are found there.

The biosphere is a vital system

LEARNING OBJECTIVE

To study why the biosphere is a vital system.

Learning outcomes

▶ To know how the biosphere provides resources for people.
▶ To understand how it is increasingly being exploited commercially for energy, water and mineral resources.

KEY TERMS

Resource – a stock or supply of something that is useful to people.

Biosphere – the part of the Earth and its atmosphere in which living organisms exist or that is capable of supporting life.

Exploitation – the act of using natural resources.

Finite resource – a resource that will eventually run out.

Water cycle – the closed system in which water moves between the atmosphere, the oceans and land.

How does the biosphere provide resources for people?

The Earth has always provided **resources** which have been used by people to survive, from the food that we eat to the materials we use to build homes and heat them. Many of the medicines that we use come from plants. Figure 6.3 gives some examples of each of these resources.

Resource	Human use
Oil	Used as a fuel to produce electricity and to power engines.
Wood	Used to as a building material and as a fuel to provide heat.
Wheat	Used to make bread and breakfast cereals.
Periwinkle	Used in medicines to treat leukaemia.

⬆ **Figure 6.3** Natural resources and how we use them.

How is the biosphere being exploited for energy, water and mineral resources?

Energy

The **biosphere** is being **exploited** in a number of ways to provide energy.

● Oil is extracted from the ground and used to power engines in many forms of transportation. The main use of oil for transportation is to power cars for private use. Other uses of oil are as a fuel to provide electricity.
● Coal has been mined for hundreds of years. It was first used as a fuel to heat people's homes and later used to produce electricity. The extraction and use of coal has changed between different countries but coal is still a major energy provider in many countries of the world.

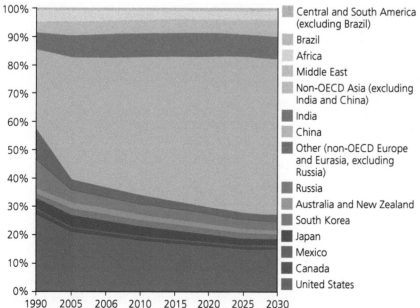

Figure 6.4 World coal consumption by region 1990–2030.

Use	Percentage
Electricity generation	68
Steel manufacture	7
Cement plants	4
Other industry	8
Heating	3
Other uses	10

⬆ **Figure 6.5** World uses of coal, 2012.

Practise your skills

Construct a pie chart to show the information in Figure 6.5.

More recently the biosphere has been exploited to provide green energy. Wind turbines have been built on land and at sea to provide energy using the power of the wind. Solar panels have been put in fields to provide electricity using solar radiation. More recently, the sea has been exploited to provide energy through the use of wave and tidal barrages.

Water

People use water in many ways. In a domestic situation water is used for drinking, washing, toilets and cleaning. It is also used in the production of electricity in thermal power stations. Farmers use water to irrigate their crops; the need for this varies depending on the location; for example in the USA 37 per cent of all water used is for irrigation. Water is also used by many different industries, from the food industry to the paper industry. In Canada, the paper industry uses 45 per cent of all water used by industry. Water is also used as a means of transporting materials and people from one place to another.

All of these uses of water have an impact on the water which is being used. The amount of water in the world is **finite** and human's use of it is interfering with the **water cycle**.

⬆ **Figure 6.6** Renewable power from the sea: wave barrages, Cardiff.

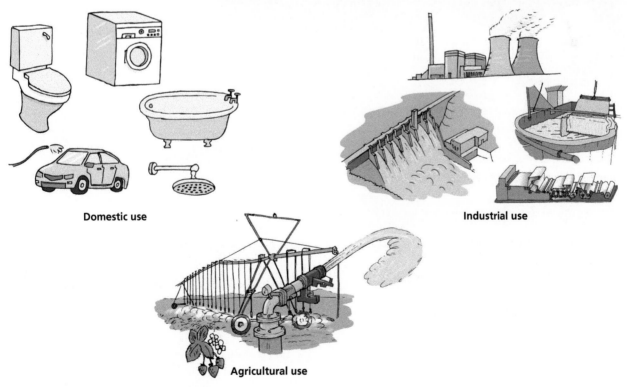

Domestic use

Industrial use

Agricultural use

⬆ **Figure 6.7** How water is being exploited.

Mineral resources

Minerals have been extracted and used by humans for thousands of years. We use them in our everyday lives without even realising it, for example when you cleaned your teeth this morning! Gold and silver are easily recognised as minerals used in the making of jewellery, but silver is also used in the making of mirrors. Figure 6.8 shows a number of minerals that are used in the construction of a house and the making of a car. It is easy to see that the use of minerals is being exploited, and that certain minerals will soon start to be in short supply.

KEY TERMS

Mineral – a solid, naturally occurring non-living substance, such as coal or diamonds.

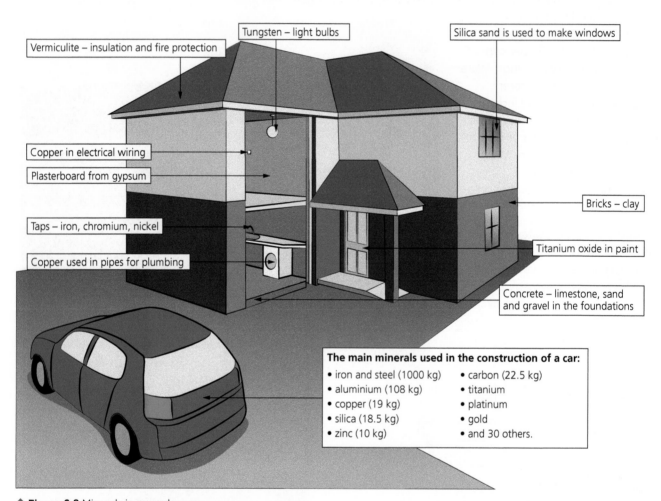

Vermiculite – insulation and fire protection

Tungsten – light bulbs

Silica sand is used to make windows

Copper in electrical wiring

Plasterboard from gypsum

Taps – iron, chromium, nickel

Copper used in pipes for plumbing

Bricks – clay

Titanium oxide in paint

Concrete – limestone, sand and gravel in the foundations

The main minerals used in the construction of a car:
- iron and steel (1000 kg)
- aluminium (108 kg)
- copper (19 kg)
- silica (18.5 kg)
- zinc (10 kg)
- carbon (22.5 kg)
- titanium
- platinum
- gold
- and 30 others.

⬆ **Figure 6.8** Minerals in everyday use.

ACTIVITIES

1 Name one resource that is both a fuel and a building material.
2 Describe how water as a resource is being exploited.
3 Explain how the energy resource is being exploited.

Extension

Research different plants that are used to treat diseases.

Review

By the end of this section you should be able to:

✓ explain how the biosphere provides resources for people
✓ understand how the biosphere is increasingly being exploited commercially for energy, water and mineral resources.

The UK has distinctive ecosystems

What is the distribution and characteristics of the UK's main ecosystems?

Type of ecosystem	England (%)	Northern Ireland (%)	Wales (%)	Scotland (%)
Farming	58	41	42	19
Moorlands and mountains	4	19	10	42
Grasslands	16	18	25	18
Woodlands	7	14	15	17
Wetlands	2	3	3	2
Urban	13	5	5	2

⬆ **Figure 6.9** Ecosystems across the UK.

Practise your skills

Construct a divided bar chart to show the information in Figure 6.9.

Marine ecosystem in Scotland most degraded in the UK!

The Firth of Clyde on the west coast of Scotland has been named as the most degraded marine ecosystem in the UK. Over 100 years of overfishing of cod, mackerel and herring, to name but a few species, has led to this sorry state of affairs. Few fish are caught by the fishing boats today as none are left in the waters.

In September 2008, in response to this sad state of affairs, a fully protected marine reserve was established in Lamlash Bay on the Isle of Arran – the first in Scottish waters. It is hoped that fish will breed in this area and then find their way back to the waters of the Firth of Clyde. If we want to continue to enjoy Great British fish and chips, we can only hope it works!

KEY TERMS

Moorland – land which is not intensively farmed. It is found in upland areas of the UK and tends to have acidic, peaty soils. The plants are small shrubs such as heather; there are few trees.

Heathland – tends to be open countryside in lowland areas. The plants are small shrubs, such as heather and gorse, with a few silver birch trees.

Deciduous – broad-leaved trees, such as oak and ash, which lose their leaves in the autumn and regrow them each spring.

Coniferous – trees which stay in leaf all year round (evergreens).

Wetlands – areas of low-lying land that is predominantly wet and boggy. Some wetland areas have been drained, such as the Somerset Levels and the Fens. The term 'wetland' also refers to small ponds and river estuaries.

Type of ecosystem	Distribution	Characteristics
Moorland ↑ **Figure 6.10** Moorland in North Yorkshire.	Found in highland areas with heavy rainfall. Examples in the UK include the Cairngorms in Scotland, North Yorkshire, the Pennines, Dartmoor, Exmoor and mid Wales.	Moorlands have been created by people. The hills used to be covered in trees and shrubs but the moorland ecosystem has developed through grazing the land with sheep and managing it for grouse shooting. In many areas the moors are burnt to control the growth of plants. Typical plants: soils are acidic and peaty so only certain plants can survive, such as bell heather and bracken. Typical animals: red deer and foxes. Typical birds: buzzards and grouse.
Heathland ↑ **Figure 6.11** Heathland in Devon.	Much of lowland UK has heaths. Heaths are found in Cornwall, Devon and Dorset, at Cannock Chase in Staffordshire and the Gower Peninsula in Wales.	Heathlands have dry sandy soils which can have depressions that are peaty and boggy. The sandy soil is free draining, acidic and has few plant nutrients. In the past heathlands were used for sheep grazing and for building materials. Typical plants: they contain small shrubs such as heather and gorse, but trees such as silver birch will colonise the area if they are not controlled. Typical animals: rabbits and hares; they are particularly important for reptiles. Typical birds: nightjar and skylark.
Woodland ↑ **Figure 6.12** Mixed woodland in Devon.	The UK has many different types of woodland. In England, Wales and Northern Ireland, native trees are broad-leaved, deciduous trees such as oak and ash. In Scotland the native tree is the Scots pine, although broad-leaved trees are also found in lowland areas. There are plantations of non-native conifers in many upland areas, such as the valleys of southern Wales and the Lake District.	Trees are the dominant plant. Broad-leaved trees tend to be deciduous, which means they lose their leaves in autumn and regrow them each spring. This leaf fall provides rich humus for the wood. Coniferous woods are made up of conifers which have needle-like leaves. They shed and replace their leaves throughout the year. Their seeds are protected by cones. Typical plants: in addition to trees, mosses and lichens grow under the canopy, as well as plants such as bluebells and ferns. Typical animals: roe deer and badger. Typical birds: sparrow hawk and tawny owl.
Wetland ↑ **Figure 6.13** The Somerset Levels.	Wetlands in the UK range from ponds and streams to rivers like the Severn and areas such as the Somerset Levels, the East Anglian Fens and the Norfolk Broads. They are also found in central and northeast Scotland, Wales and Northern Ireland.	Lowland fens have peaty, fertile soils that are periodically waterlogged. They support lush vegetation. Much of the land has been drained to use as farmland. Typical plants: reeds and bulrush grow along the sides of the streams. Typical animals: otters. Typical birds: mallard and teal.

How important a resource are marine ecosystems?

The UK's marine resources have great economic, environmental and social value. The marine area of the UK is three and a half times its land area. It is rich in resources, which were estimated to be worth £46 billion in 2005. The most recent development is the use of the marine environment for large-scale renewable energy developments.

The marine environment provides vital goods and services:

- oil, natural gas and renewable energy
- sand and gravel for the construction of roads and buildings
- seafood
- ports, through which 90 per cent of our imports and exports travel
- sport and recreation.

The UK's seas also absorb vast amounts of greenhouse gases while releasing oxygen. They moderate our climate, making it warmer in the winter than the UK would be given its latitude, and cooler in the summer, as well as giving millions of people the opportunity for leisure and recreation at the many coastal resorts around the country.

KEY TERMS

Aquaculture – breeding of fish in pens under controlled conditions.

Colonise – to become established in an area.

Resource/activity	Gross value to the UK economy (£ billion)	Number of people employed in that activity
Oil and gas	37.00	290,000
Ports	5.05	54,000
Telecom	2.70	26,750
Recreation	1.29	114,670
Fisheries	0.20	31,633
Aquaculture	0.19	
Renewable energy	0.05	4,000

⬆ **Figure 6.14** Resources from the marine environment.

Practise your skills

Construct a bar chart to show the information from the first two columns in Figure 6.14.

How are human activities causing problems for marine ecosystems?

The pressure on the UK's seas to provide resources has never been greater and this is causing problems for marine ecosystems. The problem of overfishing has been recognised for many years and laws have been introduced to address it at both a national and EU level, the latest being the creation of Marine Protected Areas (MPA). No fishing is allowed to take place in these areas of the sea.

The fastest-growing human activity is marine energy. The extraction of oil and natural gas over the last 50 years has had an impact on marine ecosystems in the North Sea. The expansion now is into renewable energy in the form of wind farms, wave farms, barrages and tidal turbines. These constructions will have a far greater effect on marine ecosystems as there are more of them and their construction requires interference with the sea bed.

↑ Figure 6.15 Wind farm in Kent.

The construction industry is also increasingly looking to the sea, rather than quarries on land, to provide sand and gravel. The growth in shipping also requires the development of port facilities.

Perhaps the largest growth area, however, is in marine leisure and the growth of marina facilities to provide recreation facilities for a growing population in the UK.

Due to these increasing and sometimes conflicting demands on marine ecosystems, the Marine and Coastal Access Act was passed in 2009. Its aim is to help with some of these concerns by managing the activities that take place in marine environments and balancing them with conservation.

Review

By the end of this section you should be able to:

✓ explain how the biosphere provides resources for people
✓ understand how the biosphere is increasingly being exploited commercially for energy, water and mineral resources
✓ describe the distribution of the UK's main ecosystems
✓ describe the characteristics of the UK's main ecosystems (moorlands, heaths, woodlands, wetlands)
✓ understand the importance to the UK of marine ecosystems as a resource
✓ explain how human activities are causing problems for marine ecosystems.

ACTIVITIES

1 Copy out and complete the following sentences:
   Cannock Chase is an example of ....
   Dartmoor is an example of ....
   The Norfolk Broads are an example of ....
2 Describe the different types of woodland found in the UK.
3 State three resources that we get from marine ecosystems.
4 Explain how human activities are causing problems for marine ecosystems.

Extension

Research the location and numbers of Marine Protected Areas around the UK. The following website will help: www.ukmpas.org/mapper.php.

Tropical rainforests have a range of distinguishing features

■ LEARNING OBJECTIVE

To study the distinguishing features of tropical rainforests.

Learning outcomes

▶ To know the biotic and abiotic characteristics of the tropical rainforest ecosystem.
▶ To be able to explain the interdependence of biotic and abiotic characteristics and the nutrient cycle.
▶ To understand why rainforests have very high biodiversity.
▶ To be able to describe how plants and animals have adapted to the rainforest environment.

What are the biotic and abiotic characteristics of the tropical rainforest ecosystem and how have plants and animals adapted to these conditions?

The **abiotic** (or non-living) characteristics of the tropical rainforest are the amount of rainfall, temperature, **soil** and light that the forest receives. Its **biotic** factors are the plants, animals and humans that can be found there. These characteristics can be seen on Figure 6.18 (page 100). The figure also contains information on how some plants and animals adapt to life in the rainforest environment.

KEY TERMS

Abiotic factors – the physical, non-living environment, such as water, wind, oxygen.

Biotic factors – the living organisms found in an area.

Detritovore – an animal that feeds on dead plant or animal matter.

Organic material – something that was once living.

Inorganic material – something that was never living matter.

Soil – the top layer of the earth in which plants grow; it contains organic and inorganic material.

Biomass – the amount or weight of living or recently living organisms in an area.

Nutrient cycle – the movement and exchange of organic and inorganic material into living matter.

Food chain – a series of steps by which energy is obtained and used by living organisms.

Food web – a network of food chains by which energy and nutrients are passed from one species to another; it is essentially 'who eats who'.

Biodiversity – the number of species present in an area.

Limiting factors – factors that limit biodiversity/population size, such as temperature, moisture, light and nutrients; these factors are in abundance in tropical rainforests, which accounts for their high biodiversity.

◀ **Figure 6.16** Plants in the shrub layer of the rainforest.

	Jan	Feb	Mar	Apr	May	Jun	Jul	Aug	Sep	Oct	Nov	Dec
Temperature (°C)	28	28	28	27	28	28	28	29	29	29	28	28
Rainfall (mm)	278	278	300	287	193	99	61	41	62	112	165	220

⬆ **Figure 6.17** Climate data for Manaus, Brazil.

The hummingbird lives in the canopy. It has strong flight muscles; figure-of-eight wing beats allow it to hover in the air.

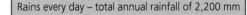

Rains every day – total annual rainfall of 2,200 mm

Toucans live in the canopy. They have long bills to reach fruit on branches that are too small to support their weight.

Temperature range between 27 °C and 30 °C

Very little light variation throughout the year – 12 hours daylight, 12 hours night

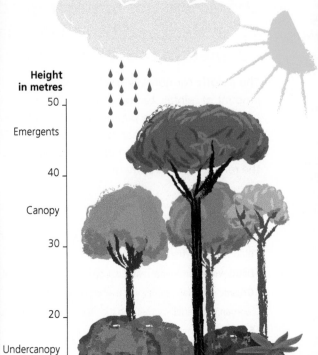

Height in metres

50 — Emergents

40 — Canopy

30 —

20 — Undercanopy

10 —

Shrub/forest floor 0 —

The trees in the canopy have small leaves to prevent water loss through transpiration

The harpy eagle lives in the canopy. It has a 2 m wing span and is so powerful that it can snatch a sloth from a tree in flight.

The large trees have buttress roots which give them stability because of their great height. The roots are also a nutrient store.

Soil poor quality; nutrients are washed down through the soil by the heavy rains. This forms a hard pan, which is a layer of solid nutrient lower down in the soil that cannot be accessed by plants.

Plants on the forest floor have adapted: they have large leaves due to lack of light, and drip tips to help them to shed rainwater quickly.

Sloths live in the canopy. They use camouflage and amazing slowness to escape predators. Green algae grows in the sloth's fur, which helps to camouflage it in the forest canopy. Sloths are among the slowest moving animals of all. They hang from branches in the canopy and are so still that predators such as jaguars don't see them. Their fur grows the other way so that the heavy rainfall runs off them easily in their upside-down position.

⬆ **Figure 6.18** The biotic and abiotic characteristics of tropical rainforests and how plants and animals have adapted to these characteristics.

What are the characteristics of the tropical rainforest food web?

Every organism needs energy to live and grow. A **food chain** is the sequence of who eats who in an ecosystem to obtain the energy to survive. A network of food chains is known as a **food web**. The food web starts with plants, which are known as producers; they gain their energy from the Sun. Plants are eaten by herbivores, or primary consumers. Primary consumers are eaten by secondary consumers, which in turn may be eaten by tertiary consumers. When an organism dies it is eaten by tiny microbes, which are known as **detritivores**. The nutrients are then recycled within the system. Figure 6.19 shows a food web in the tropical rainforest.

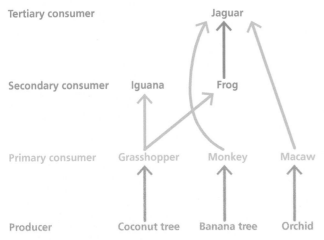

↑ **Figure 6.19** A tropical rainforest food web.

What is the nutrient cycle?

Nutrients are chemical elements and compounds that are needed for organisms to grow and live. The **nutrient cycle** is the movement of these compounds from the non-living environment to the living environment and back again. For example, a tree loses its leaves, the leaves fall to the forest floor and quickly decompose because of the hot and damp conditions. The resultant litter provides nutrients for the tree to grow. In the rainforest the majority of the nutrients are stored in the **biomass** with small amounts stored in the litter and soil. This is due to the heavy rainfall, which leaches nutrients down through the soil to an area where the plants cannot reach them. The tropical rainforest has a very high **biodiversity** because of the hot and wet conditions provided by the climate. Due to this, and consistent hours of sunlight all year round, there are very few **limiting factors**; this allows a great variety of plants to grow in the rainforest.

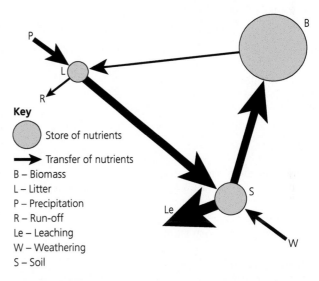

Key

○ Store of nutrients

→ Transfer of nutrients

B – Biomass
L – Litter
P – Precipitation
R – Run-off
Le – Leaching
W – Weathering
S – Soil

↑ **Figure 6.20** The nutrient cycle in the tropical rainforest (Gersmehl model).

Review

By the end of this section you should be able to:

✓ describe the biotic and abiotic characteristics of the tropical rainforest ecosystem
✓ understand the interdependence of biotic and abiotic characteristics and the nutrient cycle
✓ explain why rainforests have a very high biodiversity
✓ describe how plants and animals have adapted to the rainforest environment.

ACTIVITIES

1 How are sloths adapted to living in the rainforest?
2 State two ways that plants adapt to living in the rainforest.
3 Copy out and add to the tropical rainforest food web (see Figure 6.19). Add at least one other animal or plant to each level of the food web.

Extension

This topic has many new terms to learn. Make an ecosystem glossary and learn all the new terminology.

Tropical rainforests provide a range of goods and services, some of which are under threat

To study the range of goods and services provided by the tropical rainforest ecosystem.

Learning outcomes

▶ To know the goods and services provided by the tropical rainforest ecosystem.

▶ To be able to explain how climate change presents a threat to the structure, functioning and biodiversity of tropical rainforests.

▶ To understand economic and social causes of deforestation.

▶ To be able to describe the political and economic factors that have contributed to the sustainable management of a rainforest in Costa Rica.

KEY TERMS

Structure – the structure of a tropical rainforest is the layers of plants and animals in the forest.

Function – the function of a tropical rainforest is its ecosystem and how it works.

Transpiration – evaporation of moisture from the leaves of a plant.

Eutrophication – the growth of algae on water courses due to the increase in chemical fertilisers being used on the land.

Monoculture – the growing of one crop on large areas of land.

Which goods and services are provided by the tropical rainforest ecosystem?

The rosy periwinkle, which is found in Madagascar, has properties that can halt the progress of Hodgkin's disease for 58 per cent of sufferers. Sales of medicines from this one plant are about US$160 million a year.

The chemical structures of rainforest plants have been used as templates to make other medicines. For example, the blueprint for aspirin is derived from extracts of willow trees.

Medicines
120 prescription medicines sold globally are derived directly from rainforest plants.

Quinine, which helps to cure malaria, is an alkaloid extracted from the bark of the cinchona tree found in tropical forests in Latin America and Africa.

Bananas grow in tropical forests. They became popular in the twentieth century and are now a US$5 billion global industry.

Coffee is traditionally grown in the shade of the rainforest canopy; 25 million people globally make their living from the growth and sale of coffee.

Food stuffs

Black pepper is a spice used around the world. It grows on the flowering vine of the Piperaceae family in tropical regions of the world, especially South East Asia.

Palm oil is grown in many tropical rainforest regions of the world; it is used globally in food products such as pizza dough, biscuits and bread.

Wood from trees such as mahogany and teak is used for flooring and furniture in temperate regions, such as the USA and the UK.

Local people use wood from the rainforest for building materials and fuel. This is becoming more of a problem as population numbers in tropical areas continue to rise.

Timber

Zip wires are constructed through the canopy and under canopy for tourists to enjoy. There are also hanging bridges and courses for them to walk around.

White-water rafting: this is organised on the rivers that flow through the rainforest.

Recreation
The rainforest provides many opportunities for recreational activities. A number of countries have developed the rainforest in this way to provide an income which does not involve deforestation.

River boat rides: a good way to see the flora and fauna of the rainforest as the roads are very poor.

Nature trails on the forest floor or in the canopy are organised with trained guides who teach the tourists about the rainforest.

↑ **Figure 6.21** Goods and services provided by tropical rainforest ecosystems.

How does climate change present a threat to the structure, functioning and biodiversity of tropical rainforests?

Climate change presents a threat because the **structure** and **function** of rainforests rely on the climate. Therefore, if there are changes in temperature and rainfall distribution, the rainforest will be unable to survive in its present form. The rainforest is vulnerable to climate change because its resilience has been weakened by human activities such as deforestation. If part of the forest is felled or suffers from a fire, it has an impact. The humidity in the rainforest and the vast amount of **transpiration** means that much of the rainfall is recycled from the forest itself. If the forest is felled there are fewer trees to provide water through transpiration and therefore less rainfall occurs. Climate change means that eastern areas of the Amazon rainforest may receive twenty per cent less rainfall by 2030. This will cause the temperature to rise and will have a major impact on the forest in that area. The Amazon rainforest contains 40 per cent of the species on earth. This biodiversity will be threatened by less rainfall and higher temperatures because links in the food chain could be broken if species fail to adapt to the new climate conditions. The predictions are that deforestation and climate change could damage or destroy 60 per cent of the Amazon rainforest by 2030.

What are the economic and social causes of deforestation?

The economic causes of deforestation are related to a country using its natural resources to generate income. The social causes of deforestation relate to pressure from the growing population in many countries where there are tropical rainforests. This has resulted in more timber being felled for use as fuel and building materials, and to make way for new roads and houses.

Agriculture

⬆ **Figure 6.22** Farming in the Brazilian Amazon.

Tropical rainforest is cleared for agriculture as many believe that the soils must be rich to support the lush vegetation. The forest is felled and burned; this adds nutrients to the soil which last for a few years. The cleared area is quickly planted and good crops are harvested. After a few years large amounts of fertiliser are needed to produce high yields. This makes the farming less profitable and causes fertilisers to be washed into rivers, causing **eutrophication**. The land is then abandoned or left to cattle ranchers.

Some land on floodplains is suitable for growing cash crops on a **monoculture** system such as rice and tea. This can be risky because of disease spreading through the crop or price fluctuations on global markets.

Resource extraction

⬆ **Figure 6.23** Gold mining in the Amazon rainforest.

Resources have been extracted from rainforest areas for many years. Governments such as Brazil's have sold the rights to minerals in the rainforest as a way of raising money to help develop their country. The indigenous inhabitants are rarely consulted which has caused many disagreements because they believe that they own the land and that the government has no right to grant mineral rights to large companies. The forest has been cleared for large-scale mining operations including access roads and settlements where the workers live.

Most mining in the Brazilian Amazon today is for gold. This releases mercury into the environment, which is very dangerous to the health of both humans and other top carnivores.

Population pressure

⬆ **Figure 6.24** Population pressure in the Amazon Rainforest.

The main reason for deforestation is population pressure due to **overpopulation**. This relates to the growth in population in the country where the **tropical forest** is situated, which places growing demands on the forest area. It is predicted that the world's population will reach 8 billion by 2026. Over 99 per cent of this growth will be in developing countries, many of which have areas of rainforest.

In Brazil, the biggest increases in population are in the Amazon. Ten cities in the Brazilian Amazon doubled their population between 2000 and 2010, with the region's population increasing by 23 per cent to 25 million. This growth is shown in the city of Parauapebas which, over the past ten years, has changed from a frontier settlement with gold mines and gunfights to a sprawling urban area with an air-conditioned shopping mall. The reason for this growth is the promise of work in new energy, **resource extraction** and agricultural developments in the area. This attracts migrants from all over Brazil, which still has many people living in poverty. The growth of Parauapebas is because of employment in a large iron ore mine and the plans for more mines as demand increases from China. The population has grown from 154,000 to 220,000 since the 2010 census.

On the outskirts of the city, **favelas** of wooden shacks stretch to the horizon on land that used to be rich in tropical rainforest biodiversity.

Located example | Sustainable management of the tropical rainforest in Costa Rica

KEY TERMS

Overpopulation – too many people living in an area for the area to support.

Favela – homes for the poor in Brazil. They are made from waste materials and have no water supply, electricity or toilets. They are also usually known as shanty towns.

NGO – non-governmental organisation; a not-for-profit organisation that is not under government control. They are usually set up by private individuals and can be funded by donations or governments.

Carbon credits – a permit which allows the holder to emit one ton of carbon dioxide or another greenhouse gas; they can be traded between businesses or countries.

For example, a steel producer in the USA has been allowed to emit ten tons of CO_2 but knows it will emit eleven tons. The company could buy one credit from Costa Rica to ensure it keeps to international law. Costa Rica has many carbon credits because of its rainforest. In this way wealthy countries are encouraging poorer countries to protect their rainforests.

Ecotourism – travel to natural areas that does no damage, conserving the environment and improving the well-being of local people.

The rainforest in Costa Rica is being managed sustainably by the government in a number of ways.

- National parks and reserves take up twenty per cent of the land area. People are not kept out of these areas completely but their use is monitored and carefully managed.
- Reserves that are owned and protected by private owners make up another five per cent. For example, the Monteverde Cloud Forest Reserve.
- Non-governmental organisations (**NGOs**) and charities have been encouraged to take an interest; for example, the World Land Trust.
- Direct government action, for example, new laws.
- Selling **carbon credits** to wealthy nations in order to protect the rainforest.
- **Ecotourism** has become the main export of the country, bringing in US$1.92 billion annually.

Year	Forested area (%)
1990	50.2
1995	48.4
2000	46.5
2005	48.8
2010	51.0
2015	52.1

⬆ **Figure 6.25** Forested area in Costa Rica.

Practise your skills

1 Draw a line graph of the information in Figure 6.25.
2 Explain why a line graph is an appropriate way to display this information.

Direct government action

In 1979, Costa Rica passed legislation giving tax deductions and grants to owners of rainforest if they conserved their forest area and used it to benefit society by protecting water resources, biodiversity and scenic beauty. The government issues forest protection certificates and pays landowners US$50 annually for every hectare of forest they protect.

In 1995, the government set up national parks to protect eighteen per cent of the country's territory; privately owned reserves protect another thirteen per cent. The areas targeted for protection are those with high biodiversity.

The government has decentralised its decision making with regards to protecting the rainforest. It has grouped all protected areas in the country into eleven eco-regions. Each area is allowed to make decisions on how its rainforest will be protected.

In 1997 the Costa Rican government introduced the Certificate for Sustainable Tourism (CST) for businesses that can prove their commitment to sustainable tourism. The certificate has five levels, which include how the business looks after the environment and local people.

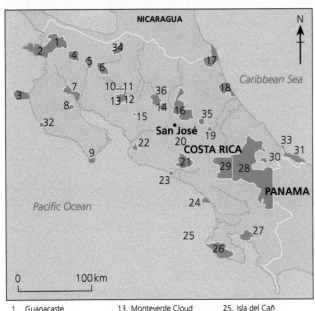

1. Guanacaste	13. Monteverde Cloud Forest Reserve	25. Isla del Cañ
2. Santa Rosa	14. Poas Volcano	26. Corcovada
3. Marino Las Baulas	15. Peñas Blancas	27. Piedras Blancas
4. Rincón de la Vieja	16. Brauilio Carrillo	28. La Amistrad
5. Miravalles Volcano	17. Barra del Colorado	29. Chirripo
6. Tenorio Volcano	18. Tortuguero	30. Hitoy Cerere
7. Palo Verde	19. Turrialba Volcano	31. Gandoca Manzanillo
8. Barra Honda	20. Tapanti	32. Ostional
9. Cabo Blanco	21. Los Quetzales	33. Cahuita
10. Arenal Volcano	22. Carara	34. Caño Negro
11. Santa Elena Cloud Forest	23. Manual Antonio	35. Irazu Volcano
12. Children Eternal Rain Forest	24. Marino Ballena	36. Juan Castro Blanco

⬆ **Figure 6.26** National Parks and reserves in Costa Rica.

NGO projects

One of these projects is run by FUNDECOR (the Central Volcanic Mountains Development Foundation). Using a US$10 million grant from USAID (aid from the USA), it works with local landowners to help them manage the rainforest sustainably so that they can qualify for government grants. In 2014, FUNDECOR was working with 400 landowners who owned approximately 46,000 hectares of land in the Central Volcanic Mountain Region.

Ecotourism

In 1983, an ecotourism project called Rara Avis S.A. started in the Braulio Carrillo National Park. The idea was to show that the rainforest is a valuable economic resource that should be protected. The reserve looks after 485 hectares of rainforest and has indirectly conserved another 1,000 hectares. The previous owners planned to cut down the trees and sell the timber but now it is being managed sustainably. It fulfils the government criteria for environmental management and social management as all of the employees are from the local village of Las Horquetas.

⬆ **Figure 6.27** Ecotourism in Costa Rica.

Private reserves: Monteverde Cloud Forest Reserve

This 10,500 hectare reserve was founded in the 1950s by a group of Americans and is now known as one of the most outstanding wildlife sanctuaries in the world. Its varied climate and altitude has a very high biodiversity. The reserve allows **ecotourism** but restricts the numbers of tourists it lets in by only having one place to stay in the reserve. It has guided walks and trails.

Charities: World Land Trust

A World Land Trust project concentrated on the Ora Peninsula, which is one of the top twenty places of highest biodiversity in the world. The project was to conserve the rainforest but also to work with local farmers. It purchased just over 2,000 hectares of land, which is now part of the Corcovado National Park. The trust worked with local farmers promoting growing vanilla to sell commercially. Small-scale ecotourism was also developed in the area.

Carbon credits

Another way that Costa Rica is using its rainforest as a **commodity** is by selling carbon credits. Wealthy countries buy them to offset the carbon emissions that they produce. This is a way for Costa Rica to earn money from its rainforest without cutting down the trees. In 1999, this idea generated US$20 million for the country.

↑ **Figure 6.28** Monteverde Cloud Forest Reserve.

Deforested areas

Deforested areas of the rainforest are also being used to support the country economically. For example, in the Talamanca region deforested land is being used for farming. Commercial crops such as bananas and cocoa are grown and sold for export.

ACTIVITIES

1 State two goods and two services provided by tropical rainforests.
2 Describe how climate change will have an impact on tropical rainforests.
3 Explain the economic causes of deforestation in tropical rainforests.
4 State two ways that the government is trying to protect the rainforest in Costa Rica.
5 What is meant by the term 'ecotourism'?
6 Explain three ways in which the rainforest provides income for Costa Rica.
7 How are deforested areas in Costa Rica being used to generate income?

Extension

Use the internet to research the Monteverde Cloud Forest Reserve (www.monteverdeinfo.com), an example of a private reserve in Costa Rica.

Practise your skills

Use GIS mapping to research the amount of rainforest destruction in Brazil over the past twenty years.

Review

By the end of this section you should be able to:

✓ describe the goods and services provided by tropical rainforests
✓ explain how climate change presents a threat to the structure, function and biodiversity of tropical rainforests
✓ understand the economic and social causes of deforestation
✓ describe political and economic factors that have contributed to the sustainable management of a rainforest in a named region.

Deciduous woodlands have a range of distinguishing features

LEARNING OBJECTIVE

To study the distinguishing features of deciduous woodlands.

Learning outcome

▶ To know the biotic and abiotic characteristics of the deciduous woodland ecosystem.
▶ To be able to explain the interdependence of biotic and abiotic characteristics and the nutrient cycle.
▶ To understand why deciduous woodlands have a moderate biodiversity.
▶ To be able to describe how plants and animals have adapted to the deciduous woodland environment.

KEY TERMS

Hibernate – to spend the winter in close quarters in a dormant (sleeping) condition.

Practise your skills

1 Draw a climate graph for the climate data given for London in Figure 6.30.
2 Compare the climate data for Manaus (on page 99) and London.

What are the biotic and abiotic characteristics of the deciduous woodland ecosystem and how have plants and animals adapted to these conditions?

The abiotic (non-living) characteristics of deciduous woodlands are the amount of rainfall, temperature, the soil and light that the forest receives. The biotic factors are the plants, animals and humans that can be found there. These characteristics can be seen in Figure 6.29. The figure also contains information on how plants and animals have adapted to life in the rainforest environment.

⬆ **Figure 6.29** Bluebells in a deciduous woodland field layer.

	Jan	Feb	Mar	Apr	May	Jun	Jul	Aug	Sep	Oct	Nov	Dec
Temperature (°C)	4	5	7	9	12	16	18	17	15	11	8	5
Rainfall (mm)	54	40	37	37	46	45	57	59	49	57	64	48

⬆ **Figure 6.30** Climate data for London.

The nightingale migrates to Africa in September, returning to the UK in April, to avoid the cold months when the woodlands offer little food.

In spring, deciduous trees grow thin, broad, lightweight leaves. These leaves capture the sunlight easily and allow the tree to grow quickly as the temperature warms and the days grow longer. However, these leaves have too much exposed surface area for the cold winter months and, therefore, the tree loses its leaves as the weather becomes colder and daylight hours shorter.

Temperature range between 4 °C and 17 °C. Long periods of light in the summer, approximately 18 hours, contrasting with short days in the winter of about 8 hours of light.

Total annual rainfall 1,000 mm.

Canopy layer – trees such as oak and ash.

Height in metres

Sub-canopy layer – trees such as rowans and dogwoods, and shrubs such as rhododendrons.

Field or herb layer – plants in this layer **flower early in the year** before the trees in the canopy have grown their leaves, which block out the light.

30

20

10

0

Ground layer – this area is dark and damp; mosses and lichens grow here.

The soil is fertile. The autumn leaf fall means that there are plenty of nutrients. Earthworms in the soil help to mix the nutrients. The tree roots are deep and therefore help to break up the rock below, which gives the soil more nutrients.

Hedgehogs hibernate during the cold winter months from about November to April.

Squirrels store food in the ground under fallen leaves so that they have food in the colder months.

⬆ **Figure 6.31** The biotic and abiotic characteristics of deciduous woodlands and how plants and animals have adapted.

What are the characteristics of the deciduous woodland food web?

Every organism needs energy to live and grow. A food chain is the sequence of who eats who in an ecosystem to obtain the energy to survive. A network of food chains is known as a food web. The food web starts with plants, known as producers, which gain their energy from the Sun. Plants are eaten by herbivores, or primary consumers. Primary consumers are eaten by secondary consumers which, in turn, may be eaten by tertiary consumers. When an organism dies, it is eaten by tiny microbes which are known as detritivores. The nutrients are then recycled within the system. Figure 6.31 shows a food web in deciduous woodland.

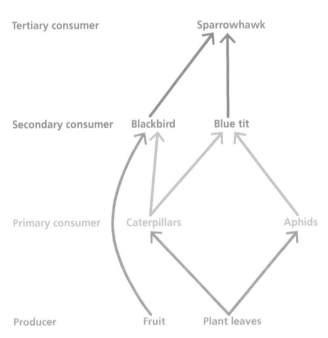

⬆ **Figure 6.32** The characteristics of the deciduous woodland food web.

What is the nutrient cycle?

Nutrients are chemical elements and compounds that are needed for organisms to grow and live. The nutrient cycle is the movement of these compounds from the non-living environment to the living environment and back again. In the deciduous woodland nutrients are stored in the biomass and soil in almost equal amounts, with a slightly smaller store in the litter. Flows of nutrients move freely between these stores.

Deciduous woodland has a moderate biodiversity because of warm summers with a consistent amount of rainfall and long hours of daylight. The autumn leaf fall supplies plenty of nutrients for the woodland plants to live and grow.

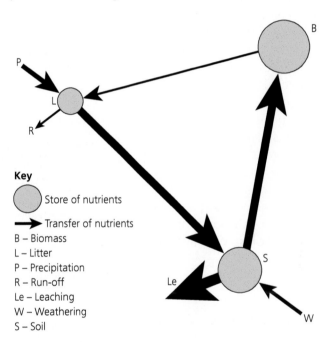

⬆ **Figure 6.33** The nutrient cycle in a deciduous woodland (Gersmehl model).

Review

By the end of this section you should be able to:

✓ know the biotic and abiotic characteristics of deciduous woodland ecosystems
✓ explain the interdependence of biotic and abiotic characteristics and the nutrient cycle
✓ understand why deciduous woodlands have moderate biodiversity
✓ describe how plants and animals have adapted to the deciduous woodland environment.

ACTIVITIES

1 State two ways that plants have adapted to life in deciduous woodlands.
2 Copy out the deciduous woodland food web (Figure 6.32) and add at least one other animal or plant to each level.
3 Describe the characteristics of the deciduous woodland nutrient cycle.

Extension

Explain what is meant by the nutrient cycle.

Deciduous woodlands provide a range of goods and services, some of which are under threat

■ LEARNING OBJECTIVE

To study the range of goods and services provided by deciduous woodland ecosystems.

Learning outcomes

▶ To know the goods and services provided by deciduous woodland ecosystems.
▶ To be able to explain how climate change presents a threat to the structure, function and biodiversity of deciduous woodland ecosystems.
▶ To understand the economic and social causes of deforestation.
▶ To be able to describe different approaches to the sustainable management of deciduous woodlands in the Wyre Forest, West Midlands.

KEY TERMS

Broad-leaved trees – deciduous trees which lose their leaves in winter, such as oak and elm.

Short rotation coppice – trees, usually willow, grown specifically to be used fuel for biomass boilers for domestic heating or power stations. They are planted densely and harvested on two- and five-year cycles.

Ancient woodlands – contain trees that were planted before 1600.

Afforestation – the planting of trees in an area that has not been forested before.

Which goods and services are provided by deciduous woodland ecosystems?

Woodlands in the UK are managed as a resource for the goods and services that they can provide. They are also managed for the benefit of the wildlife that lives within them.

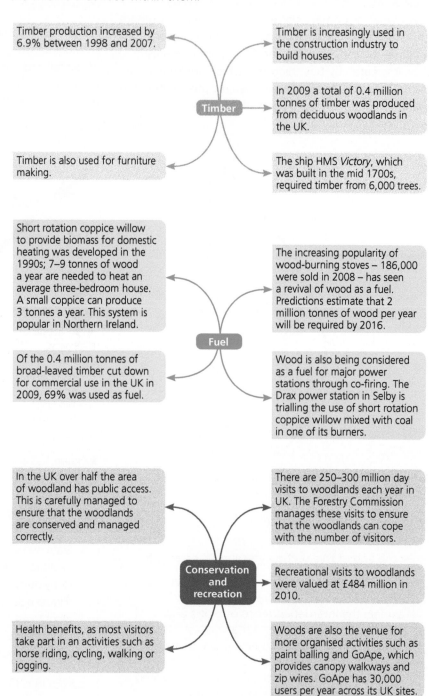

Timber production increased by 6.9% between 1998 and 2007.

Timber is increasingly used in the construction industry to build houses.

In 2009 a total of 0.4 million tonnes of timber was produced from deciduous woodlands in the UK.

Timber

Timber is also used for furniture making.

The ship HMS *Victory*, which was built in the mid 1700s, required timber from 6,000 trees.

Short rotation coppice willow to provide biomass for domestic heating was developed in the 1990s; 7–9 tonnes of wood a year are needed to heat an average three-bedroom house. A small coppice can produce 3 tonnes a year. This system is popular in Northern Ireland.

The increasing popularity of wood-burning stoves – 186,000 were sold in 2008 – has seen a revival of wood as a fuel. Predictions estimate that 2 million tonnes of wood per year will be required by 2016.

Fuel

Of the 0.4 million tonnes of broad-leaved timber cut down for commercial use in the UK in 2009, 69% was used as fuel.

Wood is also being considered as a fuel for major power stations through co-firing. The Drax power station in Selby is trialling the use of short rotation coppice willow mixed with coal in one of its burners.

In the UK over half the area of woodland has public access. This is carefully managed to ensure that the woodlands are conserved and managed correctly.

There are 250–300 million day visits to woodlands each year in UK. The Forestry Commission manages these visits to ensure that the woodlands can cope with the number of visitors.

Conservation and recreation

Recreational visits to woodlands were valued at £484 million in 2010.

Health benefits, as most visitors take part in an activities such as horse riding, cycling, walking or jogging.

Woods are also the venue for more organised activities such as paint balling and GoApe, which provides canopy walkways and zip wires. GoApe has 30,000 users per year across its UK sites.

↑ Figure 6.34 Goods and services in deciduous woodland ecosystems.

How does climate change present a threat to the structure, function and biodiversity of the deciduous forest ecosystem?

Recent climate change has not had a major impact on woodland structure and function, although small changes can be identified. Increasing temperatures have led to faster tree growth in some areas but, overall, there has been little impact. The lack of information about the impact of climate change on deciduous woodlands is partly due to the fact that the trees are long lived and can adapt to climate variability. They are, therefore, fairly resilient to the small changes in climate that have happened in recent years.

There have been more storms however, which can have an impact on the stability of the trees. In some areas increased droughts in the summer have had a detrimental effect on tree growth. It is also feared that the stress that droughts put on trees may make them more susceptible to disease. Droughts also lead to more forest fires, which have a major impact on woodland areas. In 2011, the Swinley Forest fire destroyed 200 hectares of forest.

Changes have been noted in the fauna of woodland ecosystems, such as the birds, which will eventually cause changes to the structure of the ecosystems as some species disappear and others arrive. For example, in Wytham Woods near Oxford, blue tits and great tits are breeding two weeks earlier than they did in 1980, which means they have time for more broods.

Milder winters might also cause problems as many trees need cold weather to help them to reset their clocks for spring. Without this, fruiting and flowering may be disrupted. Milder winters also mean that pests and diseases are not killed in winter frosts. It is predicted that temperatures in the UK will rise by 2.5 °C during the twenty-first century. This will mean that species will have to move north by 300 km or 300 m uphill to find the growing conditions they require or adapt to the new conditions. This would be fine for mobile species but some woodland trees might not be able to adapt fast enough. Plants such as bluebells and wood anemones are particularly under threat. This will result in the complex food web of **ancient woodlands** being disrupted by a succession of damaging impacts related to climate change.

⬆ **Figure 6.35** Walkways and zip wires in Grizedale Forest, Lake District.

What are the economic and social causes of deforestation?

Deforestation in the UK has taken place over many centuries. The first cause was an economic one: the demand for more land for agricultural purposes. Later, a social cause was population growth and the resulting demand for food; the growth of urban areas for people to live in saw the removal of yet more of the native forest. Timber extraction to generate income has been occurring for many centuries but the situation came to a head after the First World War when timber was in such short supply that only five per cent of the UK's forest remained in 1919. This led to the government setting up the Forestry Commission to promote forestry, develop **afforestation**, the production of timber and make grants available to private landowners.

Agricultural change

The farming landscape of the UK has changed many times since the Middle Ages, causing deforestation. The dissolution of the monasteries and then the agricultural revolution saw great changes to the countryside in which landowners cut down large areas of forest to plant crops. This was linked to the growing population of the time and the fact that people moved to towns and cities and so were not producing their own food.

As farming has changed through the centuries, the amount of forested area in the UK has continued to decline. Some parts of the UK have kept more of their forested area but, overall, the country's forested area declined to a low of five per cent in 1919. The impact of agricultural change on woodlands was still occurring in the 1980s as can be seen in the newspaper article below.

Deforestation in the UK!

Deforestation is happening in Kent, not just the Amazon. The new owners of a 231-hectare area of land near Maidstone in Kent have shown complete disregard for preservation orders on woodland and trees. Copses and hedgerows have been destroyed in the owner's greed for more land for intensive agriculture. Approximately 73 hectares of woodland has been lost, over half of which was ancient woodland.

Urbanisation and population growth

As the population grew in the Middle Ages there was a greater demand for housing. This meant that trees had to be felled to provide beams to support roofs. Forests were also cleared to make way for towns, especially in the north of England where the Industrial Revolution took place. This sped up of the deforestation that had been taking place in the UK for centuries. Population growth between 1945 and 1975 meant that many of the remaining deciduous woods were cut down to make way for suburbs in existing towns and cities, or for the 'new towns' that were built during that period. For example, the new town Bracknell was built by clearing large areas of Windsor Forest. Some cities, such as Birmingham and Manchester, have lost almost all of their ancient woodland. Other cities, such as London (2 500 hectares) and Sheffield (650 hectares) have retained some in parks and other woodlands, which are now used for recreational purposes.

Timber extraction

Timber has been extracted from UK forests for centuries. It was first used for house building and for fuel. The great cathedrals built across the country used thousands of trees in their construction. For example, Salisbury Cathedral required 1 000 oak trees. English oaks were also used in shipbuilding and the rise of the British Empire saw a great demand for timber in the late sixteenth century. In the 1600s the British monarchy ordered an overhaul of navy ships and about 3 000 tons of timber was felled from the New Forest each year. A survey of the New Forest in 1608 found almost 124,000 trees fit for navy timber. By 1707 that figure had fallen to less than 12,500.

Further timber extraction occurred during the First World War as it was needed to build trenches. This led to the forested area of the UK being at an all time low at the end of the war. Since 1919 the Forestry Commission has planted millions of hectares of land with fast-growing conifers. Thanks to this measure, it was able to cope with the demands of the Second World War and the demands from the developing coal industry for pit props. However, conifers are not native trees and the ecosystem of the woodlands suffered due to the dense canopy cover and the soil becoming more acidic.

Since the 1970s the Forestry Commission has changed it planting policy. As trees are extracted for timber, they are replaced with broad-leaved trees such as oak and elm in an attempt to re-establish deciduous woodlands in the UK.

More recently, the demand for timber as a fuel source has increased with the rising popularity of wood-burning stoves. Biomass boilers, which use wood chips as a fuel, have also been introduced.

Practise your skills

Use websites such as Digimap (http://digimap. edina.ac.uk) to research the amount of woodland in the UK in 1850, 1900, 1950 and 2000.

Try to discover how much of the woodland was deciduous and how much was conifer plantations in 2000.

Located example | Sustainable use and management of deciduous woodlands: Wyre Forest, West Midlands

The Wyre Forest is the largest area of ancient woodlands in England. It covers 2,400 hectares. There are also stands of conifer plantations within the forest boundary and areas of orchard, meadows and mixed farming, making a total land area of nearly 5,000 hectares. The Forest is situated in the West Midlands to the west of the Birmingham conurbation and lies on the borders of Worcestershire, Shropshire and Staffordshire. The Forest is bisected by the River Severn with Bewdley on the eastern side and Cleobury Mortimer on the west.

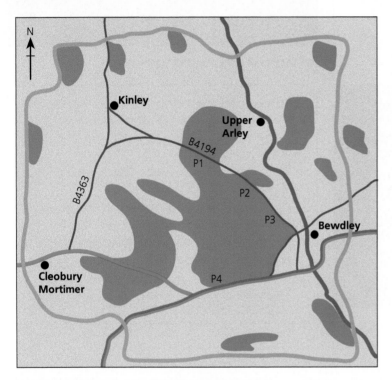

KEY

- Partnership area boundary
- Woodland areas
- River Severn
- A456
- A4117

Car parks
P1 Button Oak
P2 Hawkbatch
P3 National Nature Reserve Gateway
P4 Main visitor gateway to Wyre Forest Discovery Centre

⬆ **Figure 6.36** Sketch map of the Wyre Forest area.

Composition of the Wyre Forest	Hectares
Site of Special Scientific Interest (SSSI)	1,756
National Nature Reserve (NNR) of predominantly oak high forest of 100–120 years old	600
Ancient Semi Natural Woodland, roughly comparable to the NNR and covering the oak high forest and also the railway embankments on the current cycle path	1,564
Ancient Replanted Woodland, predominantly Forestry Commission and private ownership currently with conifer crops	1,131

⬆ **Figure 6.37** Landscapes of the Wyre Forest.

The forest is managed by the Wyre Forest Landscape Partnership (WFLP). Through meetings and consultation documents the partnership has decided on an action plan for the use and **sustainable management** of the forest. Core members of the partnership are the Forestry Commission, Natural England, Wyre Community Land Trust, Bewdley Development Trust, Cleobury Country, Wyre Forest District Council, Worcestershire County Council and English Heritage.

Woodland management

- On steep slopes deciduous woodland will be left unmanaged to develop undisturbed.
- Areas previously planted with conifers will be gradually restored to woodland with oak as the predominant tree canopy.
- Other trees such as silver birch, aspen and rowan will be encouraged with an understorey of hawthorn, hazel and holly.
- Coppice management and tree felling will be restored so that the forest develops glades of young and mature trees.
- The deer population and non-native invasive plant population will be carefully controlled.

⬆ **Figure 6.38** Woodland management.

⬆ **Figure 6.39** A woodland ride.

Wildlife management

- A varied landscape of woodland, meadows, orchards, heathland and scrub will provide a variety of habitats for wildlife in the area.
- Wildlife-rich meadows and orchards will be extended.
- Invasive species such as Himalayan balsam will be removed.
- Any cattle grazing will be carefully monitored to ensure that the wildlife habitat is enhanced, not destroyed.
- A network of woodland rides will provide corridors for wildlife as well as people.

Community management

- Local residents will be encouraged to take part in conservation work.
- There will be community woods in which local people can cut their own firewood.

⬆ **Figure 6.40** Wyre Forest centre and play area.

Leisure and recreation management

The forest will provide a place where people of all abilities can go for leisure and recreation.

- The visitor centre at Callow Hill will include displays to help people understand the forest and what it has to offer them.
- Easy bicycle access will be available from local communities.
- The Forestry Commission already provides a number of recreational activities including walking trails, cycle paths and a play area. The forest now has a GoApe experience with zip wires, Tarzan swings and walkways.

◀ **Figure 6.41** Display boards at the visitor centre.

Education

The Wyre Forest Centre has become a hub for sharing knowledge about usage and sustainable management of ancient deciduous woodlands.

- Monitoring takes place on the impact of using woodlands for recreation.
- Research is also being carried out on how the woodland is responding to external influences such as pollution and climate change.
- Children and adults from the surrounding communities, particularly from Birmingham, have been introduced to woodlands and wildlife through interactive displays and workshops.
- The Wyre Forest Landscape Partnership provides many opportunities for skills development and training, including forest industries apprenticeships and internships.

◀ **Figure 6.42** Education centre at Wyre Forest.

Review

By the end of this section you should be able to:

✓ describe the goods and services provided by deciduous woodlands
✓ explain how climate change presents a threat to the structure, function and biodiversity of deciduous woodlands
✓ understand the economic and social causes of deforestation
✓ describe different approaches to the sustainable management of deciduous woodlands in the Wyre Forest, West Midlands.

ACTIVITIES

1 State two goods and two services provided by deciduous woodlands.
2 Describe how climate change will have an impact on deciduous woodlands.
3 Explain the economic causes of deforestation in deciduous woodlands.
4 Describe two ways in which the Wyre Forest is being managed sustainably.

Extension

1 What is meant by the term 'ancient woodland'?
2 Where in the UK are other ancient woodlands located, besides the Wyre Forest?

Examination-style questions

1 Define the term food web. (1 mark)
2 Study Figure 6.2.
  a) Describe how changes in altitude affect the type of vegetation. (3 marks)
  b) Explain **one** reason for the changes describe in part (a) (2 marks)
3 Study Figure 6.20 on page 101 and Figure 6.33 on page 109.
  Compare the two nutrient cycles. (4 marks)
4 Study the OS map of Snowdonia on page 52.
  a) Identify the feature at 598545. (1 mark)
  b) Calculate the distance to the nearest km along the A4086 between the spot height 359 m in grid square 6455 and spot height 119 m in grid square 6157. (1 mark)
5 a) State **two** goods and services provided by deciduous woods. (2 marks)
  b) Explain **two** ways that plants have adapted to living in deciduous woodlands. (4 marks)
  c) Explain the economic reasons for deforestation of deciduous woodland in the UK. (4 marks)
6 Assess the following statement: 'The sustainable management of tropical rainforests is a greater concern than the sustainable management of deciduous woodlands.' (12 marks)

Total: 34 marks

Part 2 The Human Environment

In the following chapters, you will study the content you need for Component 2: The Human Environment. This component is divided into three topics:

Topic 4 Changing Cities

In this topic you will study:

Chapter 7: Changing Cities

Chapter 8: The Study of a Major UK City, Bristol

Chapter 9: The Study of a Major City, Sao Paulo, Brazil

Topic 5 Global Development

In this topic you will study **Chapter 10**, an overview of global development and a case study of development in a developing country.

Topic 6 Resource Management

In this topic you will study **Chapter 11**, an overview of resource management and one of the following two chapters:

Chapter 12: Energy Resource Management

Chapter 13: Water Resource Management

Changing Cities

Urbanisation is a global process

What were the trends in urbanisation over the past 50 years in different parts of the world?

The trends in urbanisation in the past 50 years can be seen in Figure 7.1. The fastest rates have occurred in emerging and developing countries. This is shown by the increases in the figures for Africa, Asia, Latin America and the Caribbean.

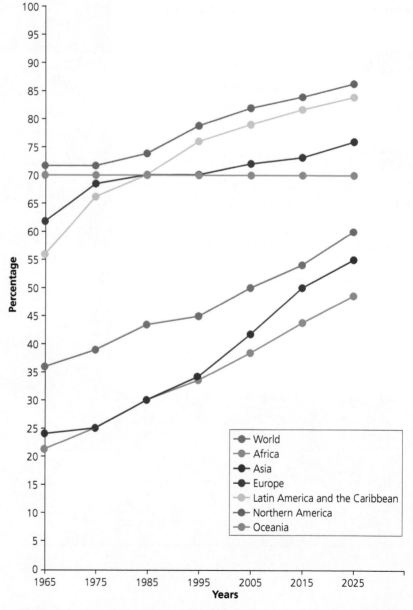

⬆ **Figure 7.1** Percentage of urban population per continent in certain areas of the world.

Year	Population (billions)
1960	3
1970	3.8
1980	4.5
1990	5.3
2000	6.0
2010	6.9
2020	7.6
2030	8.4
2040	8.9
2050	9.5

⬆ **Figure 7.2** World population growth (actual and predicted).

Practise your skills

Draw a line graph for the information given in Figure 7.2. Ensure that you indicate on your graph that information after the present year is a predicted.

KEY TERMS

Urbanisation – the increase in the number of people living in towns and cities compared to the number of people living in the countryside.

Emerging country – a country with high and medium development (HMHD).

Developing country – a country with low human development (LHD); a poor country.

Developed country – a country with very high human development (VHHD).

Natural increase – when population numbers show a positive difference between the birth rate and the death rate.

Major city – a city with a population of at least 400,000.

Human Development Index (HDI) – a measurement of life expectancy, access to education and gross national income per capita used to assess how much progress a country has made (see http://hdr.undp.org).

Rural depopulation – the movement of people from rural to urban areas.

Why has urbanisation occurred at different times and rates around the world?

Urbanisation has occurred at different times globally mainly due to the stage of development of countries.

Urbanisation in developed countries

This occurred during the nineteenth century, caused in part by the Industrial Revolution. There was a huge demand for labour in the new factories as countries industrialised, for example, in the UK. At the same time, large farming estates were enclosing land, which meant that poor villagers had less land to support themselves. People began to move in great numbers to the cities. The evidence for this can be seen in Figure 7.1, which shows that Europe and North America already had high levels of urbanisation in 1965. In the past 50 years, developed countries have continued to increase their urban areas, but at a much slower rate. This has been due to the pull of the cities, which often provide better facilities than rural areas, especially for younger generations. There are also more jobs available in urban areas, which encourages migration from rural areas.

Urbanisation in emerging and developing countries

This has occurred over the past 50 years. The evidence of this can be seen on Figure 7.1, which shows that both Africa and Asia have seen over a 50 per cent increase in the number of people living in urban areas. The main reason for this growth in urban population is the increase in population. Population growth has occurred because of decreasing death rates, which are a result of improved living conditions. More children survive past their first birthday and people are living longer. In rural areas there are not enough jobs to support the growing population so the young move to the cities looking for work. There has also been a large natural increase in the population of urban areas, partly due to the fact that many of the people who live there are of child-bearing age.

Effects of high rates of urbanisation

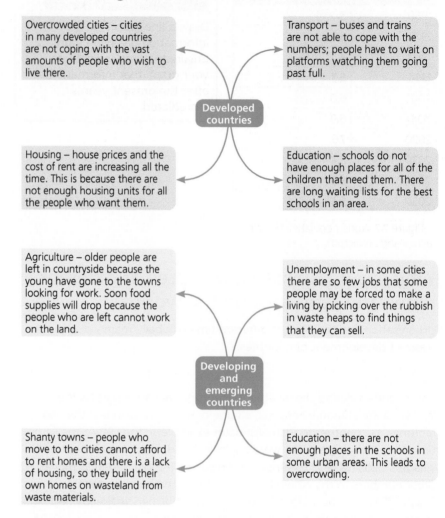

Overcrowded cities – cities in many developed countries are not coping with the vast amounts of people who wish to live there.

Transport – buses and trains are not able to cope with the numbers; people have to wait on platforms watching them going past full.

Developed countries

Housing – house prices and the cost of rent are increasing all the time. This is because there are not enough housing units for all the people who want them.

Education – schools do not have enough places for all of the children that need them. There are long waiting lists for the best schools in an area.

Agriculture – older people are left in countryside because the young have gone to the towns looking for work. Soon food supplies will drop because the people who are left cannot work on the land.

Unemployment – in some cities there are so few jobs that some people may be forced to make a living by picking over the rubbish in waste heaps to find things that they can sell.

Developing and emerging countries

Shanty towns – people who move to the cities cannot afford to rent homes and there is a lack of housing, so they build their own homes on wasteland from waste materials.

Education – there are not enough places in the schools in some urban areas. This leads to overcrowding.

⬆ **Figure 7.3** Effects of high rates of urbanisation.

Review

By the end of this section you should be able to:

✓ understand that there are contrasting trends in urbanisation over the past 50 years in different parts of the world
✓ explain how and why urbanisation has occurred at different times and rates in different parts of the world
✓ describe the effects of high rates of urbanisation.

ACTIVITIES

1. Name the three continents that have had the fastest rates of urbanisation in the past 50 years.
2. State two reasons for the high growth rate in emerging countries over the past 50 years.
3. Research one emerging country and one developing country that have experienced a high growth rate of urban areas in the past 50 years. Try to make sure that they are in different continents.

Extension

Research when the industrial revolution occurred in three different developed countries.

The degree of urbanisation varies across the UK

How is the urban population of the UK distributed and where are its major urban centres?

Figure 7.4 shows where the **major urban centres** (built-up areas) of the UK are located. It also provides details of the number of residents in some of these centres.

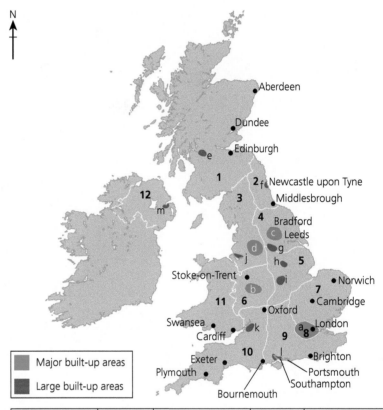

Built up areas	Letter on map	Region of the UK	Number on map	Number of residents in 2011
Greater London	a	London	8	9,800,000
West Midlands	b	West Midlands	6	2,450,000
West Yorkshire	c	Yorkshire and Humber	4	1,800,000
Greater Manchester	d	North West	3	2,550,000
Glasgow	e	Scotland	1	1,200,000
Tyneside	f	North East	2	770,000
Sheffield	g	North West	3	680,000
Nottingham	h	East Midlands	5	720,000
Leicester	i	East Midlands	5	508,000
Liverpool	j	North West	3	860,000
Bristol	k	South West	10	600,000
South Hampshire	l	South East	9	850,000
Belfast	m	Northern Ireland	12	580,000

⬆ **Figure 7.4** The UK's major urban centres (built-up areas). The table includes built-up areas with a population of over 500,000.

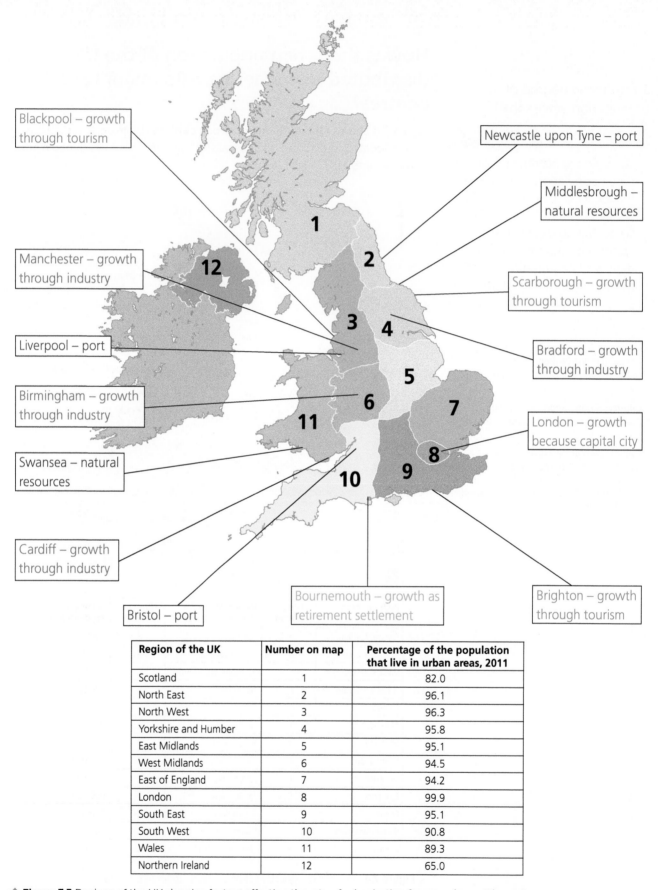

Blackpool – growth through tourism

Newcastle upon Tyne – port

Middlesbrough – natural resources

Manchester – growth through industry

Scarborough – growth through tourism

Liverpool – port

Bradford – growth through industry

Birmingham – growth through industry

London – growth because capital city

Swansea – natural resources

Cardiff – growth through industry

Bristol – port

Bournemouth – growth as retirement settlement

Brighton – growth through tourism

Region of the UK	Number on map	Percentage of the population that live in urban areas, 2011
Scotland	1	82.0
North East	2	96.1
North West	3	96.3
Yorkshire and Humber	4	95.8
East Midlands	5	95.1
West Midlands	6	94.5
East of England	7	94.2
London	8	99.9
South East	9	95.1
South West	10	90.8
Wales	11	89.3
Northern Ireland	12	65.0

⬆ **Figure 7.5** Regions of the UK showing factors affecting the rate of urbanisation for some key settlements.

What factors caused the rate and degree of urbanisation to differ between the regions of the UK?

Urbanisation took place in the UK over a long period of time, with villages becoming market towns serving a certain area. Towns that had particularly good communications would then develop into cities. Up until the mid-eighteenth century urbanisation took place at a similar rate and degree across all regions of the UK. However, during the second half of the eighteenth century the effect of the Enclosure Acts in the Midlands and north of England saw many people being forced off the land and moving to towns and cities in search of work. This was the start of the Industrial Revolution, which saw cities in the north of England develop rapidly. For example, Manchester's population increased six times between 1771 and 1831, and Bradford's population increased by 50 per cent every ten years between 1811 and 1851. Elsewhere cities such as Birmingham and London also saw enormous growth.

Growth occurred in the urban population of South Wales in Swansea and Cardiff during the nineteenth century because of the natural resources of coal, iron ore and limestone, which lead to the development of industries in this area.

Cities located on river estuaries grew as ports during the eighteenth and nineteenth centuries as trade with other countries developed. For example, in the seventeenth century Newcastle upon Tyne came after London and Bristol as the most important port in the UK. Bristol was the most important port in the UK for trade with the American colonies and the West Indies by the eighteenth century. Other ports, such as Middlesbrough, developed later as natural resources were found close to them.

Later in the twentieth century, urban population grew because of other factors. For example, North Sea oil deposits caused the growth of Aberdeen where it is piped ashore.

Growth of other coastal towns and cities occurred as they developed their tourist industry, for instance Blackpool and Scarborough in the north and Brighton in the south. The south coast towns more recently have developed as retirement settlements.

KEY TERMS

Region – a unit within a country.

Rate of urbanisation – the speed at which settlements are built.

Degree of urbanisation – the amount of built-up area that has developed in a region.

Enclosure Acts – a series of Acts of Parliament between 1750 and 1860 which stopped villages using the open fields and commons that they had been allowed to use for centuries. This meant that villagers could not make a living and had to move to find a better life. Many went to live in industrial towns.

Review

By the end of this section you should be able to:

✓ describe the distribution of the urban population in the UK and the location of its major urban centres
✓ explain the factors causing the rate and degree of urbanisation to differ between the regions of the UK.

ACTIVITIES

1. Use an outline map of the UK to mark on four major built-up areas and one large built-up area for five of the other regions shown on Figure 7.5.
2. Name one region of the UK which does not have a major built-up area.
3. State five reasons for the development of urban areas in the UK. Use examples in your answer.
4. Study Figure 7.2 on page 119. Calculate the percentage increase in the world's population between 1960 and 2000.
   a) 50%
   b) 75%
   c) 25%
   d) 100%
5. Define the term 'urbanisation'.
6. State one global trend in urbanisation. Use Figure 7.2 in your answer.
7. Explain one factor that causes urbanisation to occur at different rates in the UK.

The Study of a Major UK City, Bristol

■ LEARNING OBJECTIVE

To study how the site, situation and connectivity of Bristol influences its functions and structure.

Learning outcomes

▶ To be able to describe the site, situation and connectivity of Bristol in a national, regional and global context.
▶ To understand Bristol's structure in terms of its functions and building age.

The site, situation and connectivity of Bristol influence its functions and structure

What is the national, regional and global context of Bristol?

The national context of Bristol refers to where Bristol is compared to the rest of the UK: Bristol is in the southwest of the country, southeast of the Severn Estuary. In a regional context, Bristol is in the UK in the northwest of Europe. It is to the north of France and west of Germany. The global context means where Bristol is in the world: Bristol lies east of Canada in North America and west of Russia in Asia. These are best shown on maps (see Figure 8.1).

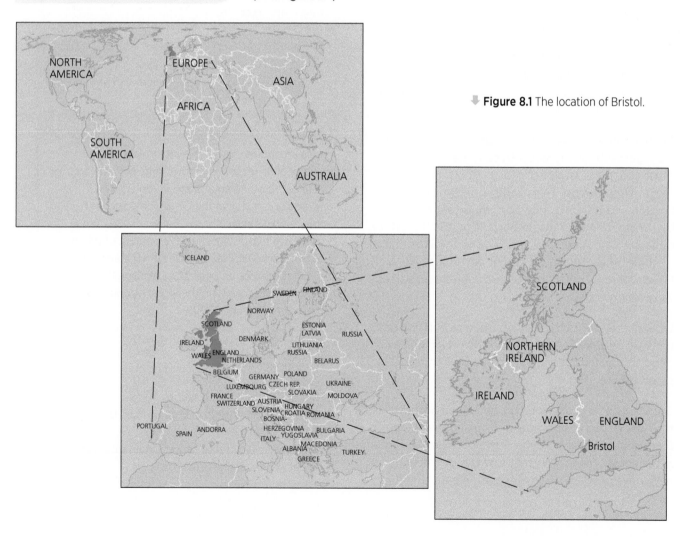

⬇ **Figure 8.1** The location of Bristol.

Site

The original settlement of Bristol grew on the confluence of the River Avon and the River Frome. The settlement then began to spread up the river valleys and the hills between them. Bristol has seven hills, which are formed from the valleys of the two rivers and their **tributaries**. These are Brandon Hill, St Michael's Hill, Old City, Redcliffe Hill, Kings Down, Clifton (or Lawrence Hill) and College Green.

The settlement grew because of trading with South Wales and Ireland thanks to the safe tidal harbour provided by the River Avon. The limestone ridge to the west of the city on which the suburb of Clifton is built is perhaps the most well-known hill. This is because the River Avon cuts through it forming the Avon Gorge over which is the Clifton suspension bridge.

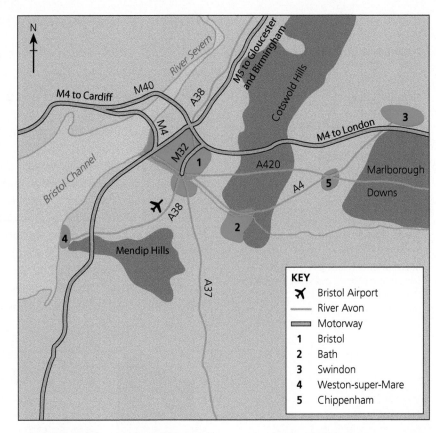

⬆ **Figure 8.2** A sketch map of Bristol showing situation and connectivity.

Situation

In a national context Bristol is situated to the southwest of the Cotswold Hills and to the north of the Mendip Hills. It is on the banks of the River Avon and 10 km east of its confluence with the River Severn. The M4 motorway is close to the northern edge of the city and the M5 is to its west. There are a number of important settlements around Bristol: Bath is 20 km to the southeast, Gloucester is 50 km northeast, Newport is 25 km northwest across the Severn Estuary, Exeter is 100 km southwest, Swindon is 50 km east and London is 150 km east.

Bristol has excellent railway links with the rest of the UK with two major stations – Bristol Temple Meads and Bristol Parkway. London, Scotland, Wales, Manchester, Birmingham and Exeter are easily accessible by train from Bristol. It also has two motorways that link it to the rest of the UK. – the M4 going east to west and the M5 going north to south. Planes fly to 112 countries from the international airport, which is to the southwest of the city.

Connectivity

Bristol is well connected. It developed as a trading settlement with Spain, Portugal and colonies in the New World. Many adventurers set off from Bristol to found new lands, for example John Cabot in the fifteenth century. In the mid-eighteenth century Bristol was England's second largest city due to its trade. Goods including sugar cane, tobacco and rum came into England through the port of Bristol. Bristol was also involved in the **slave triangle**. In the late eighteenth century rich merchants built houses in Clifton to be further from the docks.

What are the functions and building age of different parts of Bristol's structure?

Central Business District (CBD)

The **Central Business District** or **CBD** is the middle of the city. It has the tallest and oldest buildings. It is a place of renewal and redevelopment which contains the most important shops, businesses and entertainment facilities. There are important public buildings too, including the city museum and the council offices. Land in the CBD is in short supply and, therefore, is the most expensive in the urban area. As a result of this buildings tend to be tall so that they take up the minimum amount of space. Much of the centre of Bristol was rebuilt after the Second World War because the city centre was heavily bombed. Broadmead Shopping Centre was built in the 1950s to replace buildings that had been damaged by bombs (see Figure 8.3a and 8.3b). Part of this has been rebuilt recently to create The Galleries and Cabot Circus (see Figure 8.4a and 8.4b), providing the CBD with a variety of building ages and structures.

⬆ **Figure 8.3a** Broadmead Shopping Centre – redevelopment from the 1960s.

⬆ **Figure 8.3b** Broadmead Shopping Centre – the redevelopment was completed around the buildings which survived the bombing in the Second World War.

⬆ **Figure 8.4a** Cabot Circus opened in 2008.

⬆ **Figure 8.4b** The Galleries first opened in 1991 but were redeveloped in 2013.

Inner city

This is the area just outside the CBD. The function of the area is **residential** and small light industry. Much of the high-density housing was built between 1850 and 1914. The houses have no front garden and a yard or small garden at the back, usually two bedrooms and are built in terraces. Larger houses of the same age are built along the main roads, for example the A4 leading into the city. Light industry is also found in this area. Many of the larger houses have been converted into flats or used for small businesses such as dental surgeries and solicitors offices. The terraced houses were built for the workers so that they could live close to their place of work. The larger houses were built along the access roads into the city to impress visitors.

⬆ **Figure 8.5a** Larger housing on a major road leading into the city, the A4.

⬆ **Figure 8.5b** Smaller inner-city housing, Sandy Park, Bristol.

Suburbs

These can be split into inner and outer **suburbs**. The main function of the suburbs is residential; they provided housing for the workers as the settlement expanded. The inner suburbs were built between 1920 and 1940 between the two World Wars. The outer suburbs were built from the 1960s onwards.

Inner suburbs

The housing has a lower density with many of the houses being semi-detached or detached. The houses are bigger than in the **inner city**, usually having three bedrooms, a garage, and front and back gardens. The majority of the houses are owner-occupied although there are some large estates of social priority housing. Other land uses in this area are open spaces for parks, playing fields, schools and hospitals. The suburbs grew when public transport provision and private car ownership were increasing, so people could live further away from their place of work. People were also becoming wealthier and wanted larger houses with better facilities.

⬆ **Figure 8.6a** The inner suburbs: housing built along major roads.

Outer suburbs

The houses in the outer suburbs are larger and many are detached. They have larger gardens and garages. This is due to the land towards the edge of the city being cheaper. Their location on the outskirts of the city is very desirable, being away from the noise and the pollution of the CBD and inner-city area. In the 1970s housing estates were built in the outer suburbs with smaller homes to cope with the numbers of people who wished to live in the city.

⬆ **Figure 8.6b** The outer suburbs: housing estates built in the 1970s.

Urban-rural fringe

The **urban-rural fringe** is the area on the outskirts of the city. Much of this area is green belt land which was protected from development to stop the spread of the city. Over the past twenty years some development has been allowed and new housing estates and out-of-town shopping areas have been built in this area. The functions of the area are therefore to provide additional housing for a growing city and to provide shops for the residents. Transport routes have also been developed in these areas and new industrial estates on the cheaper land that is available there.

Figure 8.7b Urban-rural fringe: Cribbs Causeway, an out-of-town shopping centre.

Figure 8.7c Building on the urban-rural fringe.

Figure 8.7a Urban-rural fringe: new housing estate close to the outer ring road of Bristol.

Review

By the end of this section you should be able to:

✓ describe the site, situation and connectivity of Bristol in a national, regional and global context
✓ understand Bristol's structure in terms of its functions and building age.

ACTIVITIES

1 Copy out and complete the following sentence:
   A function of the CBD is housing / shopping.
2 What is the difference between the site and situation of a settlement?
3 Describe the site of Bristol.
4 What is the age and function of the buildings in inner-city area?

Fieldwork ideas

How has the central area of Bristol changed since the Second World War?

Primary fieldwork:

- land use of streets in the CBD
- environmental quality index
- shopping quality survey.

Secondary research:

- GOAD maps past and present to see changes to the buildings in the CBD.

The results could be shared between groups but all students should experience each of the techniques.

Bristol is being changed by movements of people, employment and services

LEARNING OBJECTIVE

To study how Bristol is being changed by movements of people, employment and services.

Learning outcomes

▶ To be able to describe the sequence of urbanisation, suburbanisation, counter-urbanisation and re-urbanisation processes and their distinctive characteristics for Bristol.

▶ To be able to explain the causes of national and international migration and their impact on different parts of Bristol.

KEY TERMS

Urbanisation – the increase in the number of people living in towns and cities compared to the number of people living in the countryside.

Suburbanisation – the growth of a town or city into the surrounding countryside, which usually joins it to villages on its outskirts making one large built-up area.

Counter-urbanisation – the movement of people from cities to countryside areas.

Re-urbanisation – the movement of people back into urban areas, usually after a city has been modernised.

Slave triangle – describes a three-part journey: ships left British ports such as Bristol with goods such as cloth and guns, they sailed to Africa where these goods were sold and enslaved people were bought. The enslaved people were taken to the Caribbean and sold. Goods such as sugar were then bought with the money and brought back to England.

Emigration – the process of moving out of a country.

Immigration – the process of moving into a country.

What is meant by the terms urbanisation, suburbanisation, counter-urbanisation and re-urbanisation, and how have these processes had an impact on Bristol?

Bristol started to urbanise when it developed as a trading settlement. This was during the fourteenth and fifteenth centuries. By the mid-eighteenth century Bristol was England's second largest city due to its trade with Spain, Portugal and colonies in the New World. Goods including sugar cane, tobacco and rum came into England through the port of Bristol. Bristol was also involved in the slave triangle. Bristol expanded in the late eighteenth century into the Clifton area when rich merchants built houses there to be further from the docks. The suburbs began to grow in the inter-war period with houses being built in areas such as Brislington. Newer suburbs such as Stockwood were built in the 1960s. Counter-urbanisation saw a decline in Bristol's population in the later half of the twentieth century but re-urbanisation has seen the development of many new housing estates and small towns built on the edge of the city from the 1980s onwards, such as Bradley Stoke to the north of Bristol.

Year	Population
1086	1,500
1200	2,000
1377	6,345
1400	9,600
1500	10,500
1600	15,000
1700	20,000
1750	45,000
1801	64,000
1851	159,945
1901	323,698
1921	367,300
1931	384,200
1941	402,000
1951	422,400
1961	425,200
1971	428,000
1981	384,800
1991	396,500
2001	380,600
2011	428,100

Practise your skills

1 Draw a line graph for the information shown in Figure 8.8.
2 Describe the information shown by your graph. Use data in your answer.

⬆ **Figure 8.8** Bristol's population growth, 1086–2011.

What are the causes of national and international migration?

National **migration** is the movement of people from one area of a country to another with the intention of staying there for at least a year. International migration is the movement of people from one country to another with the intention of staying there for at least a year. There are many reasons why people migrate, including economic, social, political and environmental. These can be split into push factors, which are all the reasons why people want to leave an area, and pull factors, which are all the factors why people want to move to an area.

Push factors	Pull factors
Natural hazards – flooding, drought	Hazard-free areas of the world
War and political conflicts – lack of safety	Political stability
Lack of jobs	Better job opportunities
Lack of facilities – education, medical, housing	Plenty of facilities
High crime rates	To be closer to family
Poverty	Good climate
Crop failure	Fertile land
Pollution	More attractive quality of life

⬆ **Figure 8.9** Reasons for migration.

What is the impact of national and international migration on different parts of Bristol?

The population of Bristol has risen by 38,000 since 2001. This is the third-highest growth rate in England. The number of people living in Bristol who were born outside of the UK has risen by over eight per cent since 2001 bringing the total to fifteen per cent of the population, of which 61 per cent arrived in the UK in the last ten years. Of this number, 69 per cent arrived in the UK when they were of working age and 30 per cent arrived as children under the age of sixteen. This indicates that people have been migrating to the city from other countries, such as Eastern European countries after their accession to the EU in 2004. More than 24,000 people came to live in Bristol between 2004 and 2009. Since 2008 there has been an increase in national migration, with people moving to Bristol from other parts of the country.

Region of birth	Number of people	% of Bristol population	% of population in England and Wales
United Kingdom	365,438	85.4	86.6
Ireland	2,900	0.7	0.7
EU member countries in March 2001	9,166	2.1	1.6
EU accession countries April 2001 to March 2011	10,520	2.5	2.0
Rest of Europe	1,656	0.4	0.5
Africa	12,858	3.0	2.3
Middle East	2,241	0.5	0.5
Asia	14,642	3.4	4.1
North America	2,166	0.5	0.4
Central America	163		
South America	1,104	0.3	0.3
Caribbean	3,809	0.9	0.5
Oceania	1,571	0.4	0.3

⬆ **Figure 8.10** Bristol residents' region of birth, 2011.

Some parts of the city have seen more migration than others. The inner-city wards of Cabot and Lawrence Hill have seen a large increase in the number of people who were born outside the UK (see Figure 8.11 for details). Almost 33 per cent of the increase in the population of Bristol has taken place in these two wards. This has caused a strain on the housing stock that is available. There has also been an increase in school children as many of the migrants are young and soon have families. This also puts a strain on doctors and hospitals. Other wards, such as Hartcliffe and Stockwood, have seen no significant increase in migration figures; in fact, they have seen decreases in their total population so there has been little impact on housing and services there.

In terms of the age structure of the city, migration appears to have increased the number of people in all age ranges but particularly in the 16–49 years group, which saw nearly a 30 per cent rise in the number of people since 2001.

Number on map	Ward name	Number on map	Ward name	Number on map	Ward name	Number on map	Ward name	Number on map	Ward name
1	Avonmouth	8	Horfield	15	Ashley	22	Lawrence Hill	29	Bedminster
2	Kingsweston	9	Redland	16	Frome Vale	23	Cabot	30	Bishopsworth
3	Henbury	10	Cotham	17	Eastville	24	Brislington East	31	Filwood
4	Westbury-on-Trym	11	Clifton East	18	Hillfields	25	Brislington West	32	Hartcliffe
5	Stoke Bishop	12	Clifton	19	Easton	26	Southville	33	Hengrove
6	Southmead	13	Bishopston	20	St George West	27	Windmill Hill	34	Stockwood
7	Henleaze	14	Lockleaze	21	St George East	28	Knowle	35	Whitchurch Park

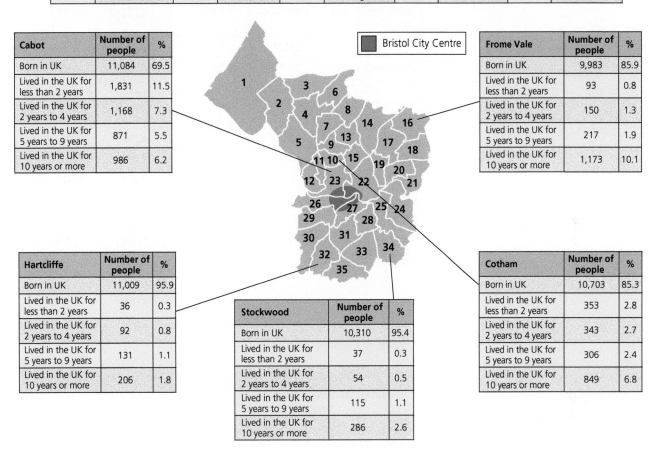

Cabot	Number of people	%
Born in UK	11,084	69.5
Lived in the UK for less than 2 years	1,831	11.5
Lived in the UK for 2 years to 4 years	1,168	7.3
Lived in the UK for 5 years to 9 years	871	5.5
Lived in the UK for 10 years or more	986	6.2

Frome Vale	Number of people	%
Born in UK	9,983	85.9
Lived in the UK for less than 2 years	93	0.8
Lived in the UK for 2 years to 4 years	150	1.3
Lived in the UK for 5 years to 9 years	217	1.9
Lived in the UK for 10 years or more	1,173	10.1

Hartcliffe	Number of people	%
Born in UK	11,009	95.9
Lived in the UK for less than 2 years	36	0.3
Lived in the UK for 2 years to 4 years	92	0.8
Lived in the UK for 5 years to 9 years	131	1.1
Lived in the UK for 10 years or more	206	1.8

Stockwood	Number of people	%
Born in UK	10,310	95.4
Lived in the UK for less than 2 years	37	0.3
Lived in the UK for 2 years to 4 years	54	0.5
Lived in the UK for 5 years to 9 years	115	1.1
Lived in the UK for 10 years or more	286	2.6

Cotham	Number of people	%
Born in UK	10,703	85.3
Lived in the UK for less than 2 years	353	2.8
Lived in the UK for 2 years to 4 years	343	2.7
Lived in the UK for 5 years to 9 years	306	2.4
Lived in the UK for 10 years or more	849	6.8

⬆ **Figure 8.11** How long people have lived in selected Bristol wards.

Figure 8.11 shows the number of people living in Bristol who have migrated from another country. The last row in each table indicates the number of people who have lived in the UK for more than ten years. The third row indicates the number of people who have lived in the UK for two years or less. These figures show the dramatic increase in the number of people living in Bristol who have recently arrived in the UK. This puts a strain on services as not only do they require employment but they also require housing, hospitals and schools for their children and, in many cases, interpreters. This puts a particular stress on schools if the children do not speak English as the teachers have to cope with this in their classrooms.

Year of arrival in the UK	Number of people
Arrived before 1941	301
Arrived 1941–50	952
Arrived 1951–60	2,594
Arrived 1961–70	4,274
Arrived 1971–80	3,451
Arrived 1981–90	4,021
Arrived 1991–2000	8,885
Arrived 2001–3	7,947
Arrived 2004–6	12,183
Arrived 2007–9	12,207
Arrived 2010–11	6,311

⬆ **Figure 8.12** Bristol residents – year of arrival in the UK.

Figure 8.13 shows the **ethnicity** of the residents of Bristol and how it has changed over a ten-year period. The numbers in all groups have increased dramatically, which is mirrored by the large increase in population in the Bristol area.

	Number of people, 2001	% of population	Number of people, 2011	% of population
All people	380,615	100.0	428,234	100.0
White British and Irish	339,406	89.1	337,283	78.8
Other white	10,124	2.7	22,309	5.2
Total black and minority groups	31,085	8.2	68,642	16.0
Mixed/multiple ethnic group	7,934	2.1	15,438	3.6
Indian	4,595	1.2	6,547	1.5
Pakistani	4,050	1.1	6,863	1.6
Bangladeshi	1,230	0.3	2,104	0.5
Chinese	2,149	0.6	3,886	0.9
Other Asian	984	0.3	4,255	1.0
Black African	2,310	0.6	12,085	2.8
Black Caribbean	5,585	1.5	6,727	1.6
Other black	936	0.2	6,922	1.6
Any other ethnic group	1,312	0.3	3,815	0.9

⬆ **Figure 8.13** Bristol's ethnicity in 2001 and 2011.

Review

By the end of this section you should be able to:

✓ describe the causes of national and international migration

✓ explain the impact of national and international migration on different parts of Bristol.

ACTIVITIES

1 Copy and complete the table below:

	Push factors	Pull factors
Economic		
Social		
Political		
Environmental		

2 Study the information in Figure 8.11 on page 131.
   a) Which wards have seen the largest growth in migrant population?
   b) What will be the impact on the quality of life for residents of these wards?
3 Which ward of Bristol would you like to live in? Give reasons for your answer.

Extension

Choose four wards in Bristol. Assess the quality of life for people living in these wards.

Globalisation and economic change create challenges for Bristol that require long-term solutions

Bristol's key population characteristics

The population of Bristol is expected to reach 500,000 by 2029. At present, it is the seventh largest city in England outside of London. Bristol Local Authority accounts for almost 70 per cent of the total population in the built-up area of the city, which is referred to as Greater Bristol. In 2011, the estimated population for this area was 617,000.

Ages	Males	Females
0–15	41,600	40,200
16–24	33,500	33,900
25–49	85,600	80,500
50–64	32,000	32,200
Over 65	25,800	32,200

⬆ **Figure 8.14** Bristol's age and gender structure, 2013.

Key characteristics of Bristol's population are:

● One in every five people living in Bristol is under the age of sixteen.
● The number of children under the age of sixteen is more than the number of people over 65.
● Bristol has a higher percentage of people in the working age range than the rest of the UK.
● People aged 20–39 make up 36 per cent of the population; the average for rest of the UK is 29 per cent.
● The highest density of population is in the inner-city wards of Cabot and Lawrence Hill.
● The wards with the highest numbers of children are found in Lawrence Hill and Filwood.
● The wards with the lowest numbers of children are Clifton and Cabot.
● The population stabilised in the 1990s and since then has grown significantly.

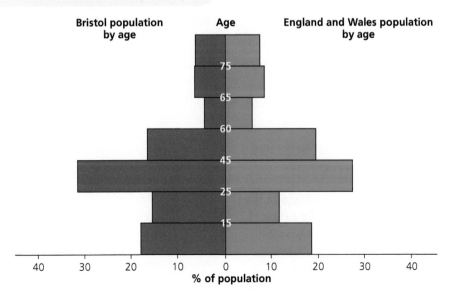

⬆ **Figure 8.15** Bristol's age structure compared to that of England and Wales.

The growth in population during the 2000s can be attributed to international migration but, since about 2009, it can be attributed to an increase in the birth rate and a decrease in the death rate in the city. During the 1980s and 1990s, the number of people living in the city went down. This mirrored a decline in population in many urban areas of the UK as people tended to settle in smaller towns and villages around cities – a trend known as counter-urbanisation. Bristol also saw many of its traditional industries closing, so people moved away from the area in search of employment.

The causes of deindustrialisation and their impacts on Bristol

Decentralisation
Many firms now have part of their production process in different parts of the world. Their headquarters are in one country and they have other parts of the company all over the world. An example is Imperial Tobacco, which still has its headquarters in Bristol but its products are manufactured in other countries.

Technological advances
Firms can have branches all over the world because of technological advances. This means that wherever the companies are located they can keep in touch with each other via the internet and can transport their products easily due to the advances in transportation systems.

Causes of deindustrialisation

Developments in transport
There have been developments in transport technology which means that goods can be moved around the world easily and quickly. This is not only the development in aircraft and containers but also the efficient motorway networks that cross Europe. This is obvious on the motorways of the UK where European lorries are very visible as they deliver goods that have been made in countries such as Poland where manufacturing costs are less. Many Marks & Spencer (M&S) products are made in Portugal, where land and labour costs are less; therefore goods can be produced more cheaply and then transported to the UK by lorry using the European motorway system. M&S also has manufacturing bases in Sri Lanka, Morocco, the South East Asia and Middle East.

Globalisation
This is the growing economic interdependency of countries worldwide which has been brought about by technological advances and development in transport systems. It has enabled countries to decentralise and has caused the deindustrialisation of many cities in the UK including Bristol.

⬆ **Figure 8.16** The causes of deindustrialisation.

A number of major companies no longer produce their products in Bristol due to deindustrialisation.

In 2010, Imperial Tobacco closed its last factory in Bristol. This is due to the fact that it could manufacture its goods more cheaply in other countries. However, its head office is still in Bristol where it employs 600 people.

In 2011, the Cadbury plant in Keynsham was closed by the US food giant Kraft which bought the company in 2010. The plant started production after the First World War and made chocolate such as Crunchie and Fry's Chocolate Cream bars for decades; 400 jobs were lost when the factory closed. The chocolate is now made in Poland and brought 1,200 miles to the UK by lorry because this is where most of the chocolate is sold.

How is economic change increasing inequality in the city and causing differences in the quality of life for its residents?

The change from an industrial-based city with a large percentage of the population working in the secondary sector to a city with a large percentage of the population employed in the tertiary and quaternary sectors has caused an increase in inequality in the city. This is because many people who worked in industries that have now closed have not been able to find new employment because they lack the skills. This has caused inequality and a decline in the quality of life for this segment of the population.

KEY TERMS

Deindustrialisation – the reduction of industrial activity in a region.

Globalisation – the way that companies, ideas and lifestyles are spread around the world.

Decentralisation – the process of spreading or dispersing power or people away from the central authority.

Edge- and out-of-town shopping – shops or facilities located away from the city centre, on the edge of cities.

Internet shopping – people buying goods online, which allows them to shop from home.

Recent changes in retailing and their impact on Bristol

There has been a decline in the CBD of many UK cities, with shops closing and many businesses have moved out of the district. The local councils including Bristol's have tried to change the trend by renting land in the CBD to development companies. These companies have developed new shopping centres, such as Cabot Circus. These new shopping centres are covered, unlike the old high streets of the CDB. They have street cafes in their large undercover shopping areas and also contain entertainment facilities and restaurants. They have good car parking facilities and excellent public transport facilities. In this way the CBD is trying to fight back against out-of-town shopping centres.

	Number of digital buyers 2014 (millions)	Number of digital buyers 2015 (millions)	Growth (%)
Germany	44.4	45.2	1.8
UK	37.5	38.2	1.9
France	29.2	29.8	2.1
Spain	17.2	18.6	8.1
Italy	14.2	15.4	8.5
Netherlands	8.3	8.5	2.4
Sweden	4.6	4.7	2.8
Denmark	3.0	3.0	0
Finland	2.6	2.7	3.8
Norway	2.5	2.6	4
Other	22.6	23.4	3.5

Figure 8.17 Number of digital buyers in western Europe, 2014–15.

Bristol does have large out-of-town shopping centres at Longwell Green to the east of the city and Cribbs Causeway to the west of the city. Cribbs has many retail parks as well as a large covered shopping centre known as the Mall, which contains all the large shops such as Marks & Spencer and Next, as well as a large John Lewis department store.

Many people now use the internet to do their shopping, both for food and luxury goods. This has also had an impact on the CBD, but people still like to 'go to the shops'. There is the idea of getting a bargain and also the convenience of being able to try things before they buy. However, the trend towards internet shopping has definitely taken some trade away from shops in the CBD. Many shops now offer a 'click-and-collect' service so that people can reserve the goods to be sure they get what they want and can also 'try before they buy'. Internet retail sales reached £52.25 billion in the UK in 2015, which was a 16.2 per cent increase on 2014. This equates to 15.2 per cent of all retail sales in the UK. In the same year, in-store sales dropped by 1.4 per cent.

Strategies aimed at making urban living more sustainable and improving the quality of life for the residents of Bristol

In 2013, a quality of living survey was completed on the twelve biggest cities in the UK. Bristol came first, with London coming seventh, but why did Bristol do so well? It is partly due to the strategies that the local council have put in place but also due to beautiful area it is in and the great variety of people and environments within the city itself.

The council measures whether it is achieving its goals by asking the residents to complete a 'Quality of life in your neighbourhood survey' annually. This is a comprehensive survey that asks a range of questions on different topics, ranging from health and education to the amount of graffiti in the area and access to leisure facilities. There are also questions on cycling to work and fear of crime. The information is available to all on the council's website and is one way that the council can measure its achievements towards its sustainable goal. The council takes sustainability very seriously and has a whole department dedicated to making it successful. It is also funding a research project with the University of the West of England, which is in north Bristol, looking at sustainable living.

Recycling

Bristol Council provides different bins for waste in the city and has done so for many years. It also has a good provision of **recycling** centres spread around the city for people to take their waste to. The council also provides a collection service for large bulky items for a fee of £15 for up to three items. As a result Bristol has one of the highest recycling rates of any city in the UK: in 2012 the residents of Bristol recycled 50 per cent of their waste. This has steadily increased from a rate of twelve per cent in 2004. This is due to the council providing home owners with kerbside recycling, which takes all of their recyclable waste. This saves residents having to travel to the waste disposal plants and therefore encourages people to recycle more.

The council has recently entered into a new deal for waste collection. It should save the council £2.5 million a year and allows residents to recycle more items such as plastics and Tetra Paks, further increasing the kerbside recycling rate.

It has also developed new ways of dealing with its untreated waste so that no waste goes to landfill. The new waste treatment plant built at Avonmouth by New Earth Solutions can deal with 200,000 tonnes of waste per year, which would otherwise have gone to landfill. It also provides electricity for approximately 13,000 homes.

Employment

The unemployment rate in Bristol is eight per cent, which is one of the lowest rates in the country. This is due partly to the work of the council, which has been active in attracting companies to the city. The council

has promised that Bristol will be the most sustainable city in the UK by 2020. This has attracted many green companies to the city. Bristol also has the highest growth of disposable income in the UK, with an average salary of £22,293, which is above the UK average of £21,473. This means that the population who live in Bristol have a high disposable income which supports job creation in service industries.

Education and health

As part of the council's 20:20 Plan, a promise has been made to improve education standards in the city's schools and to provide better health care. The council is funding projects about healthy eating in schools. This has improved people's awareness, for example, the importance of eating five portions of fruit and vegetables a day.

Transport strategies

Walking

Bristol Council has a partnership with walkit.com, a website that provides easy-to-read maps between any two points in the city. It displays the journey distance, the walking time at different paces, the amount of calories used and the amount of CO_2 saved. This will have benefits for people's health as well as cutting down on carbon emissions as less car journeys are made.

Public transport

Most of the major roads around the city now have bus lanes that cannot be used by private vehicles. Regular buses go into the city centre and the council is working towards producing a system very like the Oyster card system in London. This will enable people to use the same travel card on buses and trains in the city.

⬆ **Figure 8.18** Bus lane on the A4 leading into Bristol.

Car sharing

Bristol Council has set up a car club so that you can hire a car nearby whenever you need one. There is also a car share page on the council's website: you don't have to own a car to car share, just join and find people who are driving the same way as you to work and offer to car share with them.

Some employers have a system for car sharing within their companies; in many cases, car sharers are given priority parking (known as car club parking spaces).

Bristol has 2+ people lanes for cars. This means that, at certain times during the morning and evening rush hour, only cars with two passengers can use these lanes on the road. This encourages people to car share on their way to work.

⬆ **Figure 8.19a** Car sharing schemes: parking space reserved for car club members.

⬆ **Figure 8.19b** Car sharing schemes: car sharing lane on A4174.

Cycling

Bristol has many cycle routes and became the UK's first cycling city in 2008. The government gave Bristol Council £11.4 million to create dedicated cycle lanes, better facilities for bike users and more training for children. It created a dedicated cycleway that links the suburbs with the city centre as well as providing facilities for people who choose to cycle to work, with 300 cycle parking spaces in the city centre. There is also a scheme which repairs bikes and provides them free of charge to deprived communities. The city also has an on-street cycle hire system.

Some companies in the city, for example Network Rail, allow employees to buy a bike by paying for it monthly through their salaries; after two years they can pay off the loan and own the bike or start a new scheme. In this way the employee has a bike without the large initial cost and does not have to pay the VAT. The council has recently invested another £35 million as part of its plan to get fifth of all commuters on their bikes by 2020. Segregated cycle paths will rise from nine per cent of the road area to twenty per cent, and there will be commuter corridors heading north, east, northwest and south from the city centre.

⬆ **Figure 8.20** Bike racks in the CBD.

Affordable and energy-efficient housing

Houses are responsible for 25 per cent of the UK's carbon footprint. It is therefore important that Bristol Council does what it can to improve energy efficiency in housing in order to be more sustainable and to improve the quality of life for its residents.

Grants are available for loft insulation. The council is working with British Gas to ensure that all homes have a sufficient level of insulation to make them energy efficient.

All new developments need to submit a sustainable energy strategy to the planning committee before they can get planning permission. This should take into consideration the use of renewable energies to provide heat and light and the use of insulation to ensure that the property is built to the right specification to reduce heat loss.

The council realises that housing in Bristol is expensive and that many people cannot afford to buy or rent a property. They provide affordable housing where rents are 80 per cent of the local market rent and others where the rent is worked out on a national social formula. The homes are available through HomeChoice Bristol which provides contact details of a number of private housing associations as well as the council's housing scheme.

The council also runs help-to-buy schemes with shared ownership and rental schemes. They also provide sheltered housing for older people and retirement housing. These are all strategies that help to improve the quality of life for the people who live in Bristol.

Review

By the end of this section you should be able to:

✓ describe the key population characteristics of Bristol and reasons for population growth or decline
✓ describe the causes of deindustrialisation and their impacts on Bristol
✓ explain how economic change is increasing inequality in Bristol and the differences in quality of life
✓ explain recent changes in retailing and their impact on Bristol
✓ describe the range of possible strategies aimed at making urban living more sustainable and improving the quality of life for the people who live in Bristol.

ACTIVITIES

1 Which wards of Bristol have:
    a) the most children
    b) the least children?
2 Which age range in Bristol has the most people?
    a) 0–15
    b) 16–24
    c) 25–49
    d) 50–64
3 Give two reasons why chocolate is no longer made at Keynsham.
4 Calculate the percentage increase in recycling in Bristol between 2004 and 2012.
5 Describe the ways in which the council has tried to stop people driving their cars to work.

Extension

Assess the strategies which are used by Bristol Council to improve the quality of life for people who live in the city.

The Study of a Major City, Sao Paulo, Brazil

The site, situation and connectivity of Sao Paulo influence its functions and structure

What is the national, regional and global context of Sao Paulo, Brazil?

The national context of Sao Paulo is where it is compared with the rest of the Brazil: Sao Paulo is in the southeast of the country. The regional context of Sao Paulo is information about how it is located compared to other countries in South America: Brazil is in the centre, east of South America with Paraguay to the west. The global context is where Brazil is in the world and how it is connected to other continents and countries: Brazil is in South America; it lies east of the Pacific Ocean and west of Africa, North America is to the north of Brazil.

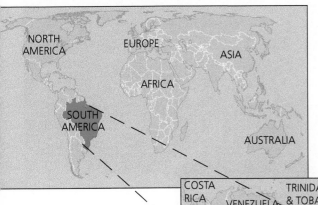

▼ **Figure 9.1** The location of Sao Paulo Brazil.

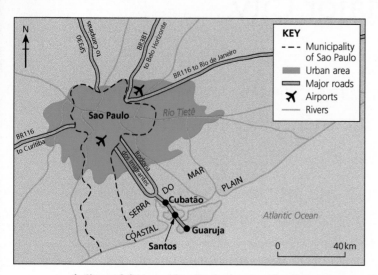

↑ **Figure 9.2** Map of Sao Paulo showing situation and connectivity.

Site

The **site** of Sao Paulo is a hilly plateau over which flow a number of rivers. The Sao Paulo city area is divided in two by the Anhangabaú River which now flows underground. The Tietê River still flows through the city. Other rivers in the area are the Tamanduateí and Pinheiros. The city is approximately 820 m above sea level.

Situation

In a national context, Sao Paulo is situated in the southeast of Brazil. It is 70 km inland from the Atlantic Ocean and is 350 km southwest of Rio de Janeiro and 330 km northeast of Curitiba. Brasília, the capital of Brazil, is 1,000 km north of Sao Paulo. Sao Paulo is on a plateau to the northeast of the coastal range of mountains known as the Serra do Mar. The main port in the state of Sao Paulo is called Santos; it is 70 km to the southeast.

Connectivity

Sao Paulo is well connected with the rest of Brazil with many of the roads and railways in southern Brazil converging on the city. There is a motorway (the Rodovia dos Imigrantes, see Figure 9.3) and a railway which link Sao Paulo to the port of Santos, 70 km to the southeast. The city has internal links with a subway system and overground train lines; it also has 16,000 buses. Two major international airports link it with the rest of the world.

↑ **Figure 9.3** The Rodovia dos Imigrantes.

KEY TERMS

Paulistanos – the name given to the residents of Sao Paulo.

Site – the land that the settlement is built upon.

Situation – where the settlement is compared to the physical and human features around it.

Connectivity – the way that the city is connected or linked to other settlements in the country as well as to other countries in the world.

Residential – an area used for housing

Public buildings – buildings owned by the council that serve the residents of the city, such as a library.

Sao Paulo metropolitan area – the whole of the built-up area; it includes Sao Paulo and a number of nearby cities; it has approximately 19 million inhabitants.

Sao Paulo city area – the inner built-up area of Sao Paulo, which has approximately 11 million inhabitants.

Peripheria – the outer edge of the city.

Cortiços – inner-city accommodation for those living in poverty. Families live in one room with shared toilets and cooking facilities. Many of the buildings were previously offices blocks or the homes of the wealthy before they left the inner-city area.

Favela – homes for those living in poverty which can be found anywhere in the city. They are made from waste materials and have no water supply, electricity or toilets.

Boulevards – wide, tree-lined streets.

What are the functions and building age of different parts of Sao Paulo's structure?

Central Business District (CBD)

⬆ **Figure 9.4** Sao Paulo city centre from Congonhas Airport, five miles away.

The CBD of Sao Paulo has both business and residential functions. Many of the buildings were constructed in the nineteenth century but, during the rapid industrialisation in the twentieth century, many high-rise buildings were built both as office blocks and as homes for the wealthy. The city centre of Sao Paulo is split into Centro Velho and Centro Novo by the valley of the Anhangabaú River. Since the first half of twentieth century the Centro Velho has developed as the centre for the financial sector, home to the headquarters of both domestic and foreign banks as well as the stock exchange. The Centro Novo is more focused on shopping, hotels and cultural establishments with museums, theatres and restaurants. It has wide boulevards and many high-rise residential blocks for wealthy residents. It also contains the main railway stations.

The CBD has seen a number of changes to its functions over the past 50 years. Firstly, wealthy residents started to move out to areas in the **inner city** such as the Jardins. This is an expensive residential area to the southwest of the CBD; this was due to the lack of space and increasing traffic in the CBD. More recently, wealthy residents and some service industries have moved to the **suburbs** and urban-rural fringe. Some people living in poverty have moved into the vacated residential and office buildings, usually without permission; these are known as cortiços. There has also been an increase in the number of street traders in the CBD, which the wealthy see as dangerous; the people who live in the cortiços see it as a way of earning a living.

Inner city

Around the CBD is the inner-city area. The main function of this area is residential but there is also some industry. This is where the migrants settled in the nineteenth and early twentieth centuries in areas such as Liberdade and Bela Vista. Bela Vista is the area of the city that is home to many migrants from Italy. Liberdade is where many migrants from Japan settled. This area also contains industry and the first favelas were established here. Many of the favelas were removed by the council but some do remain, such as Paraisópolis, where almost 43,000 people live crammed into an area of 150 hectares.

Suburbs

The suburbs of Sao Paulo have mainly residential and commercial functions. They show great contrasts between the housing. There are many favelas in these areas but also areas of very expensive housing. The area of Morumbi, a suburb of Sao Paulo in the southwest of the city is an area of high-security housing complexes with a number of parks and shopping centres, for example, Cidade Jardim Mall. There are also good hospitals and schools. The state government headquarters and the university are also in this suburb. It also contains a large number of favelas, such as Paraisópolis, which have been built on any land that is available. The Santo Amaro suburb contains the headquarters of banks and multinational companies that have moved out of the CBD. It also contains many wealthy residential areas with schools and shopping centres such as Chacara Flora, schools, and is close to Congonhas Airport. Here too there are favelas on any land that has been left vacant.

⬆ **Figure 9.5** Bela Vista, an inner-city area of Sao Paulo.

Urban-rural fringe

Many favelas sprung up in the **urban-rural fringe** in the 1980s due to the rapid population growth and lack of housing in the city. The residential usage of the urban-rural fringe has changed over the last twenty years with the development of gated communities. These are almost like mini-cities on the outskirts of Sao Paulo where wealthy residents have moved to because of the perception of a better, cleaner life on the edge of the city. Many use helicopters to commute to work in the CBD. The rich live, and many of them work, in these gated communities. See Figure 9.6 for information about Alphaville, a gated community in the northwestern part of Sao Paulo.

Gated communities

Alphaville is a gated community in the rural-urban fringe of Sao Paulo. It is surrounded by high walls and patrolled by over 1,000 armed guards. It has approximately 34,000 residents and over 2,300 businesses focused on the tertiary sector of industry. It has its own shopping centre, entertainment venues, restaurants, hospital, schools and leisure facilities.

⬆ **Figure 9.6** Alphaville, a gated community in Sao Paulo's rural-urban fringe.

ACTIVITIES

1 Copy out and complete the following sentence. A function of the CBD in Sao Paulo is ....
2 What is the difference between the site and situation of a settlement?
3 Describe the site of Sao Paulo.
4 What is the age and function of the settlements on the urban-rural fringe?

Extension

Create a fact file on Brazil. Include information such as the birth rate, death rate, gross national income, life expectancy, literacy rates and internet users.

Review

By the end of this section you should be able to:

✓ describe the site, situation and connectivity of Sao Paulo in a national, regional and global context
✓ understand Sao Paulo's structure in terms of its functions and building age.

The character of Sao Paulo is influenced by its fast rate of growth

What are the reasons for the past and present trends in Sao Paulo's population growth?

There are a number of reasons for the rapid growth of Sao Paulo: national and international migration, high rates of natural increase, economic investment and growth.

National and international migration

Sao Paulo has experienced national migration, especially from the northeast of Brazil. These migrants were attracted by the fast economic growth of the city and the promise of jobs. Much of the city of Sao Paulo was built by migrants from the northeastern states of Brazil. However, the rate of migration has decreased since the beginning of the twenty-first century.

Sao Paulo has also experienced many flows of international migration during the nineteenth and twentieth centuries. Even today, ten people move to Sao Paulo every hour. The first settlers were Portuguese but the main ethnic group in the city now is Italians. There are European, Arab, Asian, African, Jewish, Latin American and North American people living in Sao Paulo, all drawn to the city by the high rates of economic growth that it experienced in the 1950s and 1960s. Now one-fifth of the city's population is from overseas and, of that number, half of the people are Italian.

Push factors	Pull factors
In Brazil, 31 per cent of rural households have no land. They have to rent land or find work as labourers and, as farms become more mechanised, there is the risk of losing their jobs. There is little to keep people in rural areas so they move to the cities in search of work.	Infant mortality is lower in the favelas of Sao Paulo, where it is 82 per 1,000, than in the rural areas, where it is 175 per 1,000.
In the 1950s and 1960s there was a shortage of labour in Sao Paulo due to rapid economic growth of 226 per cent. Advertising campaigns were run in the rural areas to attract workers to the city.	Word sent back to the villages by successful migrants makes life in the cities seem much better than it actually is.
Bahia in northern Brazil is an area where many people live in poverty and periodically suffers from drought. It has been estimated that 3.2 million people in the state of Bahia suffer from chronic malnutrition.	The rural dwellers have high expectations of a better quality of life in the city. There are more schools and doctors as the government puts more money into services for urban areas.
Land in rural areas has been taken from the subsistence farmers who were renting it from large landowners. These landowners now want to use the land to grow cash crops such as coffee and orange juice. Just eighteen landowners control an area six times the size of Belgium.	Migration from rural areas has slowed down in Brazil, although there is still migration between urban areas for better job prospects and higher education.

⬆ **Figure 9.7** Reasons for national migration

High rates of natural increase

The growth in population in Sao Paulo over the past twenty years has been caused by a high natural increase in population. The birth rate is still high in some parts of the city and the death rate has declined with improvements in health care, diet and housing conditions. The growth rate has, however, started to slow down, from 5 per cent in 1975 to 1.3 per cent in 2013.

The causes of national and international migration

The causes of national and international migration are dealt with in Chapter 8 on page 124.

What is the impact of national and international migration on different parts of Sao Paulo?

There is a great variety of ethnic groups in Sao Paulo, which makes it a very culturally diverse place to live. The different ethnic groups can be found all over the city but certain areas are known for certain ethnic populations. For example, Bela Vista is a mixed ethnic neighbourhood with people from Portuguese, African-Brazilian, Spanish, German and English descent but, since the early twentieth century, it has been known for its large community of Italians. The neighbourhood of Liberdade has the largest Japanese community in the world outside of Japan. Since 1974 the entrance to the area has been through a *torii*, or large Japanese arch (see Figure 9.9). Today Liberdade is also home to many Chinese and Korean migrants.

Year	Population (millions)
1950	2.5
1960	4
1970	7.5
1980	12
1990	15
2000	16.5
2010	19
2020	22

⬆ **Figure 9.8** Sao Paulo metropolitan area population, 1950–2020.

Practise your skills

1 Draw a line graph for the data in Figure 9.8.
2 Why is a line graph the most appropriate form of display for this data?

⬇ **Figure 9.9** Japanese *torii* on the Rua Galvao Bueno in Liberdade.

The large number of migrants has caused a young age structure which in turn has resulted in the high birth rate. The large number of migrants has also put pressure on housing and other services such as hospitals. In Sao Paulo many of the migrants live in the favelas or cortiços which have developed since the 1980s. Twenty per cent of the residents of Sao Paulo now live in this kind of housing due to acute housing shortages.

Ethnicity	Population in Sao Paulo
Italian	6 million
Portuguese	3 million
African	1.7 million
Arabic	1 million
Japanese	665,000
German	400,000
French	250,000
Greek	150,000
Chinese	120,000
Bolivian	60,000
Korean	50,000
Jewish	40,000

⬆ **Figure 9.11** Sao Paulo metropolitan area: population by ethnic groups, 2010.

⬆ **Figure 9.10** Cortiços and favelas in Sao Paulo.

Why is the growth of Sao Paulo accompanied by increasing differences in the quality of life of its residents?

Sao Paulo is a relatively wealthy city in comparison with other cities in Brazil but poverty, unemployment and inequalities in the quality of life are still huge problems. The process of deindustrialisation, which occurred in Bristol in the twentieth century, has been happening in Sao Paulo since the 1980s causing a bigger gap between rich and poor people in the city. This has caused an increase in the unemployment rate, which is at its highest for many years.

In 2002 a report by the Sao Paulo city administration measured each district's quality of life using the United Nations Human Development Index. It found that Moema, the city's richest district, had a higher standard of living than Portugal, whereas the poorest district, Marsilac, is poorer than Sierra Leone. This inequality in wealth is partly due to the rate at which the city has grown. The growth of the population has been so rapid that the city has been unable to build sufficient housing for all the people. Therefore, the people who have jobs or have been in the city for a longer period of time tend to be the ones that are better off.

Many wealthy **Paulistanos** now live on the outskirts of the city with the centre of the city having a higher concentration of people living in poverty. The rich move around the city by helicopter – there are more than 200 helipads in Sao Paulo whereas Los Angeles has 70. In the city centre, older homes and factories have been turned into cortiços where whole families share one room which may have no electricity or plumbing. It is estimated that 600,000 people now live in the corticis. Unlike cities in the developed world, rich and poor people sometimes live next to each other, separated by walls and other security measures as shown in Figure 9.12.

In 2003 the inequalities in housing in Sao Paulo led to the occupation of abandoned high-rise blocks by 4,000 homeless people in protest against the way that the government seems to ignore homelessness.

⬇ **Figure 9.12** Extremes of wealth and poverty in Sao Paulo, Paraisópolis favela and Morumbi.

ACTIVITIES

1 International migration is caused by wars. Is this statement true or false? Give a reason for your answer.
2 What are the causes of national migration in Brazil?
3 Name one area of Sao Paulo that has a concentration of a particular migrant group.
4 Why did economic growth cause the population of Sao Paulo to increase?
5 Why has the growth of Sao Paulo caused differences in the quality of life for its residents?

Extension

1 Why did many people migrate from the northeast of Brazil to Sao Paulo?
2 Why does having a high migrant population increase the birth rate?

Review

By the end of this section you should be able to:

✓ understand the reasons for past and present trends in population growth
✓ explain the causes of national and international migration and the impact on different parts of Sao Paulo
✓ explain how the growth of Sao Paulo is accompanied by increasing inequality and the reasons for differences in the quality of life of its residents.

Rapid growth within Sao Paulo results in a number of challenges that need to be managed

To study how rapid growth within Sao Paulo results in a number of challenges that need to be managed.

Learning outcomes

▶ To understand the effects of Sao Paulo's rapid urbanisation.

▶ To be able to describe the advantages and disadvantages of both bottom-up and top-down projects to solving Sao Paulo's problems and improving the quality of life for its people.

▶ To be able to explain the role of government policies in improving the quality of life (social, economic and environmental) in Sao Paulo.

KEY TERMS

Top-down approaches – this is when the government improves an area and expects people to move into the housing they have provided. Sometimes the government borrows large sums of money from other countries to pay for the scheme.

Bottom-up approaches – these are self-help schemes. The residents of an area are in charge of what happens. They are usually given monetary help and advice on how to improve their houses.

What are the effects of rapid urbanisation?

Rapid urbanisation in Sao Paulo did not cause the problems of inequality, but it did make them much worse. At present it is estimated that twenty per cent of the population live in favelas, and that 70 per cent of the housing in the city's area is substandard.

Favelas
At the beginning of the twentieth century Sao Paulo was socially divided between the affluent who lived in the central districts and the poor who were concentrated on the floodplains of the rivers and along the railway lines. At this time Sao Paulo did not really have favelas. These started to develop in the 1930s as the city started to experience population growth. Rapid industrialisation in the1980s caused the growth of favelas across the city due to acute housing shortages. The areas did not have proper sewerage systems – much of the sewerage runs down the streets into the rivers. People access water from stand pipes which serve hundreds of people.

Unemployment
Sao Paulo could not provide jobs for all of its migrants, which led to high unemployment rates of 19% in 1998; it had reduced to 11% in 2012.

Effects of rapid urbanisation

Traffic congestion and pollution
The residents of Sao Paulo own 6.2 million cars and there are 16,000 buses on the road. At times there can be hundreds of kilometres of gridlocked roads. So many vehicles on the roads causes pollution.

⬆ **Figure 9.13** Effects of rapid urbanisation

What are the advantages and disadvantages of top-down and bottom-up approaches to solving Sao Paulo's problems?

Top-down projects are those that are instigated by the government. In some cases the government almost imposed new housing on the inhabitants of an area.

Cingapura Housing Project

This was implemented by the municipality of Sao Paulo between 1995 and 2001. The scheme was supposed to build 100,000 new homes but, in the end, only 14,000 were built. The project removed some favelas to clear land for new homes while the favela inhabitants lived in barrack-like accommodation. The new homes were built in blocks about ten storeys high. The favela residents were then expected to pay a rent of about US$26 a month for their new apartment.

Advantages:

- The new housing had clean water supply and proper sanitation.
- The new housing was built on the same land as the favelas, so people did not have to leave the area they knew.
- Leisure areas were included in the developments.

Disadvantages:

- Many favela owners have never paid rent and can't afford to.
- Favelas were demolished to build the new blocks.
- There was no provision for small businesses.
- The type of accommodation is forced on the inhabitants who have no say in what is being built.
- The living space in each apartment was very small.

Bottom-up projects are when the community are in charge of what happens. The government often provides money for the people who live in the favelas to improve their own homes. This became the policy in Sao Paulo from 2000 onwards.

Self-help scheme in Santo Andre

The scheme got together a number of different organisations to work together to improve the infrastructure and services of the area. Some of the improvements include:

- Community health projects have made health care more available.
- Literacy courses have been made available for adults.
- Recreational facilities have been made available.
- Many of the favelas have been upgraded, see Figure 9.15.
- Credit facilities have been made available to small-scale entrepreneurs so that they can expand their businesses.

Advantages:

- The community are included in the decisions that are made.
- The housing in the area will be the same type of housing but will be more substantial and will have services.
- The improvements are not just housing but help further with improving the quality of life of people in the area.

⬆ **Figure 9.14** Cingapura scheme homes.

⬆ **Figure 9.15** Self-help housing in Sao Paulo.

Disadvantages:

- The schemes take a long time to instigated.
- With so many different people involved it is hard to get agreement on how the money available should be spent.
- It is difficult to get people to accept help with the literacy programmes.

Self-help schemes in Monte Azul and Favela Jardim Jacqueline

During the 1980s the inhabitants of some of the favelas started their own self-help (bottom-up) schemes.

Favela Monte Azul is next to a polluted stream (see Figure 9.17). It is home to 3,800 people living in 400 huts that stretch up the hillside from the stream. It is situated in the southern suburbs of Sao Paulo. The self-help scheme was started by Ute Craemer, a German teacher. Its first project was to clean up the stream to provide fresh water through pipelines and to provide sanitation. In 1985, a wooden clinic was built; it was soon rebuilt as a three-storey brick house. The scheme now organises day nurseries, schools, workshops and a bakery. It has 120 volunteer helpers.

Favela Jardim Jacqueline is situated next to one of Sao Paulo's wealthiest areas (see Figure 9.18) yet its residents still experience severe cases of malnutrition. In 1994, 190 families a month were receiving baskets of food organised by a committee of nine people who begged for food and money to get the project started. The committee then turned to building a day-care centre for the favela's children who otherwise roam the streets. It is hoped that it will eventually employ eighteen staff and look after 240 children.

What is the role of government policies in improving the quality of life in Sao Paulo?

The government of has attempted to improve the quality of life for the people who live in Sao Paulo by a number of ways.

- A government bank (BNH) has funded housing projects and provided low-interest loans to lower- and middle-income people to help them to buy a home.
- A scheme which built houses for teachers and other people who worked for the government.
- A scheme to build government-owned housing which also funded self-help projects (*mutirões*) to upgrade housing in the favelas.

Other ways that the government has tried to improve the quality of life include providing an underground train system, the Metrô. It was built in the 1960s and opened in 1974.

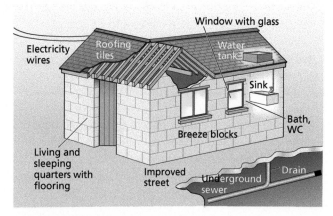

⬆ **Figure 9.16** A self-help home.

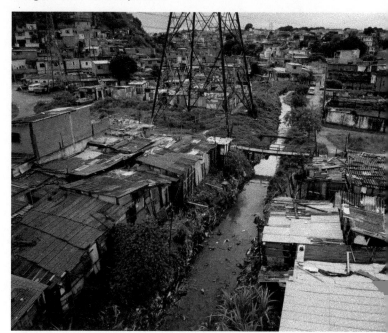

⬆ **Figure 9.17** A polluted river in Sao Paulo.

⬆ **Figure 9.18** Jardins (gardens) in Sao Paulo.

The system cuts down on traffic congestion and pollution. There are now six lines and 65 stations carrying three million passengers a day. It is an affordable and efficient way for people to move around the city but it is very crowded. Further extensions are planned.

The city has also instigated busways (similar to bus lanes in the UK) in an attempt to deal with traffic problems. Buses have sole use of these lanes but they do not cover the whole city yet – in some places buses have to merge with the other traffic, slowing the journey down.

↑ Figure 9.19 The 9 de Julho Busway in Sao Paulo.

Review

By the end of this section you should be able to:

✓ understand the effects of Sao Paulo's rapid urbanisation
✓ describe the advantages and disadvantages of both bottom-up and top-down approaches to solving Sao Paulo's problems and improving the quality of life for its people
✓ explain the role of government policies in improving the quality of life in Sao Paulo.

ACTIVITIES

1 State two ways in which the Sao Paulo Metrô improves the quality of life for people who live in the city.
2 What are the advantages and disadvantages of the self-help schemes of the 1980s?
3 Assess the different approaches to the housing problem in Sao Paulo.

Examination-style questions

1 a) Study Figure 7.1. State which area of the world had the greatest increase in urban area between 1965 and 2015. (1 mark)
   b) Define the term urbanisation. (1 mark)
   c) State **two** effects of rapid urbanisation in developing or emerging countries. (2 marks)
2 Study Figure 8.3. It shows the CBD of Bristol.
   a) Define the term CBD. (1 mark)
   b) Identify **two** pieces of evidence that show that buildings in the CBD were built at different times. (2 marks)
   c) Suggest **three** functions of the CBD. (3 marks)
3 Suggest **two** possible impacts on a city if there is a large increase in migration. (2 marks)
4 Explain how transport in developed world cities can be made more sustainable. (3 marks)
5 Study Figure 9.2.
   a) State in which direction you would travel from Santos to Sao Paulo. (1 mark)
   b) Suggest **two** reasons why a city developed in this location. (4 marks)
6 Study Figure 9.8.
   a) State how many more people live in Sao Paulo in 2010 than in 1960. (1 mark)
   b) Calculate the percentage increase in population between 1980 and 1990.
       A 12%   B 20%   C 25%   D 15% (1 mark)
7 Evaluate the role of government policies in improving the quality of life for people who live in Sao Paulo. (8 marks)

Total: 30 marks

Global Development

To study how definitions of development vary, as do attempts to measure it.

Learning outcomes

▶ To know contrasting ways of defining development using economic criteria and broader social and political measures.
▶ To understand that different factors contribute to the human development of a country: economic, social, technological and cultural, as well as food and water security.
▶ To know how development is measured in different ways: gross domestic product (GDP) per capita, the Human Development Index (HDI), measures of inequality and indices of political corruption.

KEY TERMS

Development – an improvement in the quality of life for the population of a country.

Primary sector – extractive industries such as farming, fishing, forestry and mining. Developing countries have high numbers of people employed in this sector.

Secondary sector – manufacturing industries; the number of people employed in this sector increases as a country develops.

Tertiary sector – service industries and jobs such as teaching; few people are employed in this sector in a developing country.

Quaternary sector – financial services and telecommunications.

Development gap - the difference between the parts of the world that have wealth and the parts that do not.

North–South divide – a virtual socioeconomic and political line on the globe which splits the developed and wealthy countries in the 'North' from the poorer developing countries in the 'South'.

Definitions of development vary, as do attempts to measure it

There are contrasting ways of defining development, using economic criteria and broader social and political measures

There are a number of different ways to define development. It can relate to economic, social, political or even cultural changes.

Economic development	An increase in a country's wealth. This could be an increase in people working in the secondary sector and a decrease in the numbers of people working in the primary sector. It could be indicated by a greater use of natural resources, for instance, energy use per head of population increases.
Social development	A number of changes that have a direct impact on the population's quality of life. This could include improved levels of literacy through greater access to education, better housing conditions and more doctors.
Political development	Freedom for the people to have a greater say in who governs their country.
Cultural development	This could involve better equality for all and good relations between people from different ethnic groups.

⬆ **Figure 10.1** Four types of development.

What factors contribute to the human development of a country?

The development of a country is affected by a number of factors including economic, social, cultural, technological, as well as food and water security.

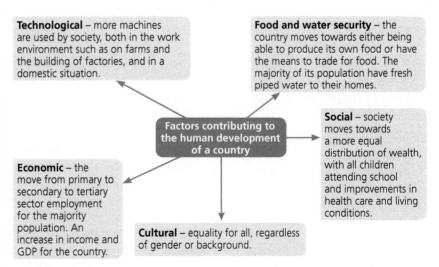

Technological – more machines are used by society, both in the work environment such as on farms and the building of factories, and in a domestic situation.

Food and water security – the country moves towards either being able to produce its own food or have the means to trade for food. The majority of its population have fresh piped water to their homes.

Factors contributing to the human development of a country

Social – society moves towards a more equal distribution of wealth, with all children attending school and improvements in health care and living conditions.

Economic – the move from primary to secondary to tertiary sector employment for the majority population. An increase in income and GDP for the country.

Cultural – equality for all, regardless of gender or background.

⬆ **Figure 10.2** Factors contributing to the human development of a country.

Development is measured in different ways

The following are all ways that the development of a country can be measured.

- **Gross domestic product (GDP)**: the value of all the goods and services produced in a country during a year, in US dollars. Per capita means that the figure is divided by the number of people who live in the country to give an average per person.
- **Human Development Index (HDI)**: this is a comparative measure of different aspects of life between countries. The measures used are life expectancy, education and standards of living (see http://hdr.undp.org).
- **Measures of inequality**: these are ways of measuring how equal people are within a country or between countries. This is often a measurement of the wealth or health care of people in a country or between countries.

- **Corruption Perceptions Index**: This is the perceived corruption in governments and the public sector. It is a perception because corruption is hidden and therefore difficult to measure. It means that government officials are using development for their own betterment rather than the betterment of the country. In the 2014 index, 100 is the perfect score so these countries have very 'clean' public sectors; 0 indicates a very corrupt public sector (see Figure 10.3). For example, Zimbabwe, currently governed by Robert Mugabe, is perceived to be very corrupt: Zimbabwe scores 21 and is ranked 157 out of 177 countries.

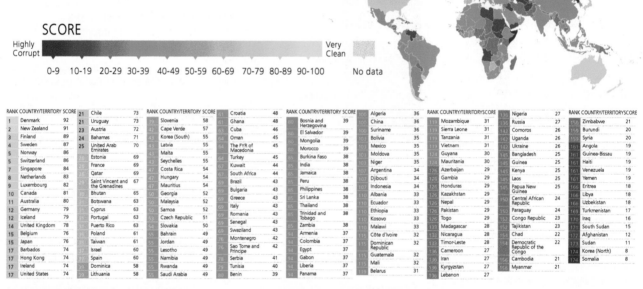

SCORE

Highly Corrupt — Very Clean

0-9 10-19 20-29 30-39 40-49 50-59 60-69 70-79 80-89 90-100   No data

RANK	COUNTRY/TERRITORY	SCORE
1	Denmark	92
2	New Zealand	91
3	Finland	89
4	Sweden	87
5	Norway	86
5	Switzerland	86
7	Singapore	84
8	Netherlands	83
9	Luxembourg	82
10	Canada	81
11	Australia	80
12	Germany	79
12	Iceland	79
14	United Kingdom	78
15	Belgium	76
15	Japan	76
17	Barbados	74
17	Hong Kong	74
17	Ireland	74
17	United States	74

RANK	COUNTRY/TERRITORY	SCORE
21	Chile	73
21	Uruguay	73
23	Austria	72
24	Bahamas	71
25	United Arab Emirates	70
	Estonia	69
	France	69
	Qatar	69
	Saint Vincent and the Grenadines	67
	Bhutan	65
	Botswana	63
	Cyprus	63
	Portugal	63
	Puerto Rico	63
	Poland	61
	Taiwan	61
	Israel	60
	Spain	60
	Dominica	58
	Lithuania	58

RANK	COUNTRY/TERRITORY	SCORE
	Slovenia	58
	Cape Verde	57
	Korea (South)	55
	Latvia	55
	Malta	55
	Seychelles	55
	Costa Rica	54
	Hungary	54
	Mauritius	54
	Georgia	52
	Malaysia	52
	Samoa	52
	Czech Republic	51
	Slovakia	50
	Bahrain	49
	Jordan	49
	Lesotho	49
	Namibia	49
	Rwanda	49
	Saudi Arabia	49

RANK	COUNTRY/TERRITORY	SCORE
	Croatia	48
	Ghana	48
	Cuba	46
	Oman	45
	The FYR of Macedonia	45
	Turkey	45
	Kuwait	44
	India	44
	South Africa	44
	Brazil	43
	Bulgaria	43
	Greece	43
	Italy	43
	Romania	43
	Senegal	43
	Swaziland	43
	Montenegro	42
	Sao Tome and Principe	42
	Serbia	41
	Tunisia	40
	Benin	39

RANK	COUNTRY/TERRITORY	SCORE
	Bosnia and Herzegovina	39
	El Salvador	39
	Mongolia	39
	Morocco	39
	Burkina Faso	38
	India	38
	Jamaica	38
	Peru	38
	Philippines	38
	Sri Lanka	38
	Thailand	38
	Trinidad and Tobago	38
	Zambia	38
	Armenia	37
	Colombia	37
	Egypt	37
	Gabon	37
	Liberia	37
	Panama	37

RANK	COUNTRY/TERRITORY	SCORE
100	Algeria	36
100	China	36
100	Suriname	36
103	Bolivia	35
103	Mexico	35
103	Moldova	35
103	Niger	35
107	Argentina	34
107	Djibouti	34
107	Indonesia	34
110	Albania	33
110	Ecuador	33
110	Ethiopia	33
110	Kosovo	33
110	Malawi	33
115	Côte d'Ivoire	32
115	Dominican Republic	32
115	Guatemala	32
115	Mali	32
	Belarus	31

RANK	COUNTRY/TERRITORY	SCORE
119	Mozambique	31
119	Sierra Leone	31
119	Tanzania	31
119	Vietnam	31
124	Guyana	30
124	Mauritania	30
126	Azerbaijan	29
126	Gambia	29
126	Honduras	29
126	Kazakhstan	29
126	Nepal	29
126	Pakistan	29
126	Togo	29
133	Madagascar	28
133	Nicaragua	28
133	Timor-Leste	28
136	Cameroon	27
136	Iran	27
136	Kyrgyzstan	27
136	Lebanon	27

RANK	COUNTRY/TERRITORY	SCORE
136	Nigeria	27
136	Russia	27
142	Comoros	26
142	Uganda	26
142	Ukraine	26
145	Bangladesh	25
145	Guinea	25
145	Kenya	25
145	Laos	25
145	Papua New Guinea	25
150	Central African Republic	24
150	Paraguay	24
152	Congo Republic	23
152	Tajikistan	23
154	Chad	22
154	Democratic Republic of the Congo	22
	Cambodia	21
	Myanmar	21

RANK	COUNTRY/TERRITORY	SCORE
156	Zimbabwe	21
159	Burundi	20
159	Syria	20
161	Angola	19
161	Guinea-Bissau	19
161	Haiti	19
161	Venezuela	19
161	Yemen	19
166	Eritrea	18
166	Libya	18
166	Uzbekistan	18
169	Turkmenistan	17
170	Iraq	16
171	South Sudan	15
172	Afghanistan	12
173	Sudan	11
174	Korea (North)	8
174	Somalia	8

⬆ **Figure 10.3** Corruption Perceptions Index, 2014. Source: Transparency International.

Review

By the end of this section you should be able to:

- ✓ define development using economic criteria and broader social and political measures
- ✓ understand factors that contribute to the human development of a country
- ✓ describe how development is measured in different ways.

ACTIVITIES

1 Give two different definitions of development.
2 Describe three factors that contribute to the development of a country.
3 Describe two ways in which development can be measured.
4 Research which three countries are perceived to be the most corrupt countries in the world using the website of Transparency International (www.transparency.org).

Extension

Comment on the statement: 'Cultural development could include greater equality for women.'

The level of development varies globally

The global pattern of development and its unevenness between and within countries, including the UK

The maps in Figures 10.5 and 10.6 on page 154 show the difference between using HDI and GNP to measure development. Some countries are placed in the top percentage in one index but not in the other. For example, Sweden and France are in the top group for GNI but do not rank at the top for HDI.

When individual countries are considered, most have areas which are richer than other areas. This is also true in towns and cities. For example, the Bronx in New York has very deprived areas but many people consider New York to be a wealthy city with a high HDI. Using broad country figures (as the two maps in Figures 10.5 and 10.6 do) hides a lot of problems within countries.

Using the UK as an example, the country is split between the north and the south. Figure 10.4 clearly shows this, with the income in the south and east being far higher than that in Yorkshire and Lancashire. This still hides the true picture however, because not all of the people who live in the south and east earn a high income. This makes their apparent poverty all the more striking. This is the same in all countries and cities throughout the world.

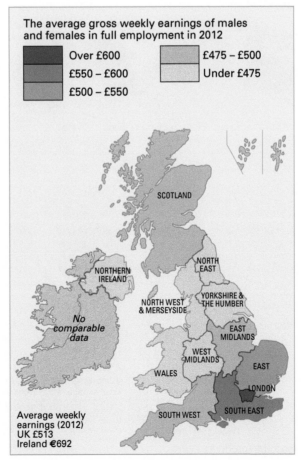

The average gross weekly earnings of males and females in full employment in 2012

Over £600
£550 – £600
£500 – £550
£475 – £500
Under £475

Average weekly earnings (2012)
UK £513
Ireland €692

◀ **Figure 10.4** UK income.

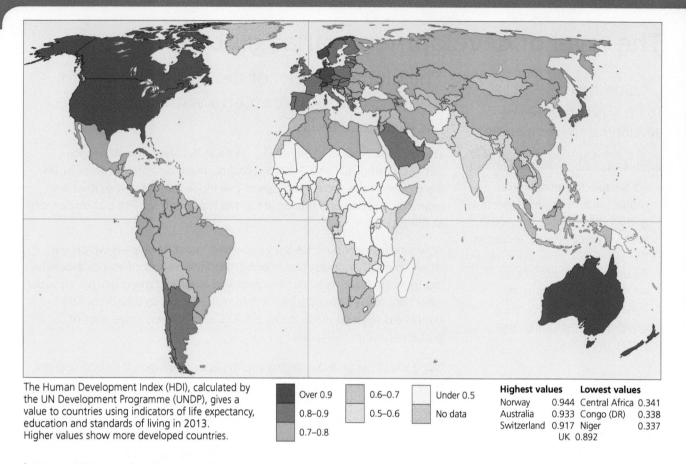

The Human Development Index (HDI), calculated by the UN Development Programme (UNDP), gives a value to countries using indicators of life expectancy, education and standards of living in 2013. Higher values show more developed countries.

■	Over 0.9	▦	0.6–0.7	□ Under 0.5
▦	0.8–0.9	▦	0.5–0.6	▦ No data
▦	0.7–0.8			

Highest values		Lowest values	
Norway	0.944	Central Africa	0.341
Australia	0.933	Congo (DR)	0.338
Switzerland	0.917	Niger	0.337
	UK 0.892		

⬆ **Figure 10.5** Human Development Index.

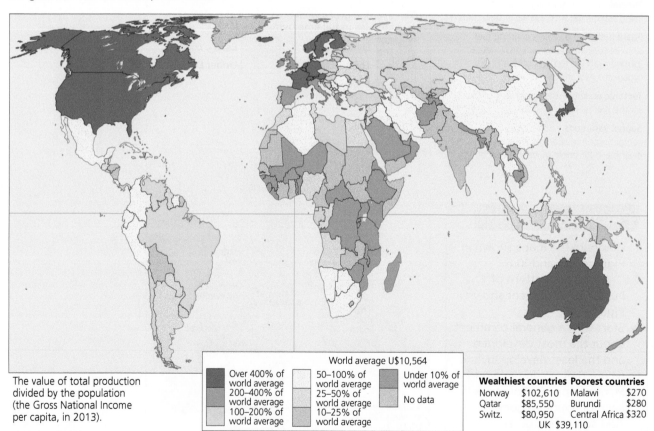

The value of total production divided by the population (the Gross National Income per capita, in 2013).

World average U$10,564

■	Over 400% of world average	□	50–100% of world average	▦	Under 10% of world average
▦	200–400% of world average	▦	25–50% of world average	▦	No data
▦	100–200% of world average	▦	10–25% of world average		

Wealthiest countries		Poorest countries	
Norway	$102,610	Malawi	$270
Qatar	$85,550	Burundi	$280
Switz.	$80,950	Central Africa	$320
	UK $39,110		

⬆ **Figure 10.6** Gross national income (product).

Factors that have led to spatial variations in the level of development globally

Climate
Countries that have average rainfall and moderate temperatures are able to support their populations with the food that they produce. In Africa, many countries suffer from frequent droughts. This means that crops die and people do not have enough food to eat. Certain diseases also thrive in hot climates, such a malaria and yellow fever; these diseases make people very weak and unable to work.

Landlocked countries
Countries that do not have a coastline find it difficult to trade their goods. They have to rely on the goodwill of their neighbours to allow them to transport their products to the coast and for them to receive imported goods.

Physical

Natural resources
Resources such as minerals and fossil fuels help a country to develop. The extraction and sale of these resources will bring income into the country.

Natural hazards
Floods, tectonic activity, droughts and hurricanes are more likely to occur in some countries than others. Many of the countries that suffer from these hazards are developing because income has to be diverted to help recover from these events on a regular basis.

⬆ **Figure 10.7** Physical factors.

Colonies
Colonies supplied food to the country that colonised them. For example, Brazil sent food and minerals to Portugal.
This hindered the development of the colony but aided the development of the colonising country.

Trade
Many trading partnerships go back to colonial times. Countries with good trading partners or countries on trade routes developed more quickly than countries that did not trade with other countries.

Historic

Politics
Countries with stable governments developed more quickly. If countries are at war or are suffering from civil wars, their income is spent on military weapons rather than on development. Also, development can be halted if a country is corrupt as the money may be spent on an affluent lifestyle for the elite group of people who rule the country.

⬆ **Figure 10.8** Historic factors.

World trade
The developing countries sell primary products to developed countries. Manufactured goods are worth more money than primary products so developed countries earn more from their trade than developing countries.

Economic

Foreign investment
This can help a country to develop because it brings money into a country. Africa is home to 15% of the world's population and receives 5% of direct international investment; Europe is home to 7% of the world's population and receives 45% of direct international investment. However, things are changing as companies from developed countries start to invest in emerging countries, for example, Coca-Cola in India.

Infrastructure
The country's roads, railways and facilities, such as electricity. Developed countries have a good infrastructure and therefore companies want to invest in them because they know their goods will be produced and moved quickly.

⬆ **Figure 10.9** Economic factors.

Factors that have led to spatial variations in the level of development within the UK

The factors mentioned above can also be applied to the different areas of the UK to explain the difference in development and wealth within the country. The table in Figure 10.10 gives some of the reasons for the UK's uneven development (see also Chapter 7, pages 118–123).

Reason		Development outcome
Physical	Relief	The south of the UK is flatter; this aids development as urban areas can be easily built upon. The north and west are more mountainous, making urban areas and communication routes more difficult to build.
	Climate	The south and east of the UK have a better climate than the rest of the country with less rainfall. This makes it a pleasanter areas to live in.
	Natural resources	The Midlands and North and South Wales started to develop with the discovery of natural resources; in the first instance this was the mining of coal.
	Position	The south and east of the country are closer to the communication links to Europe. This makes companies want to locate in this area.
Historical	Politics	The seat of the government is in London, in the South East. This made it a highly desirable location for business in the past as they were close to where decisions were being made and found out about them quickly.
	Colonies	Although ships sailed for the colonies from ports on the west of the country, all of the decisions were taken in London on the east of the country.
Economics	Infrastructure	The infrastructure in the London area is the best in the country. All roads lead to the centre of London. Companies who located there would be able to trade with the rest of the country easily.
	Foreign investment	Most international investment into the UK is in London although the government has tried to encourage foreign firms to invest elsewhere, for example Honda in Swindon.

⬆ **Figure 10.10** Reasons for the UK's uneven development.

Review

By the end of this section you should be able to:

✓ describe the global pattern of development and its unevenness between and within countries, including the UK
✓ understand the factors that have led to spatial variations in the level of development globally and within the UK.

ACTIVITIES

1 Study Figure 10.5 and Figure 10.6.
  Name two countries that are in the top percentages of the world's earnings but not in the top categories for the HDI.
2 List the physical factors that have hampered the development of some countries.
3 Describe three reasons for the UK's uneven development.

Extension

What is the development gap?

Uneven global development has had a range of consequences

LEARNING OBJECTIVE

To study how uneven global development has had a range of consequences.

Learning outcomes

▶ To be able to describe the impact of uneven development on the quality of life in different parts of the world.

KEY TERMS

Literacy – the ability to read and write.

Employment structure – the numbers of people employed in each sector of industry.

Anomaly – something that is outside the norm.

The impact of uneven development on the quality of life in different parts of the world

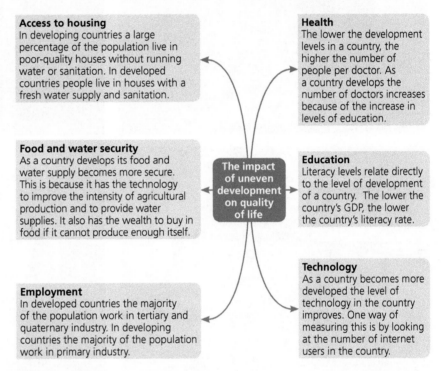

Access to housing
In developing countries a large percentage of the population live in poor-quality houses without running water or sanitation. In developed countries people live in houses with a fresh water supply and sanitation.

Health
The lower the development levels in a country, the higher the number of people per doctor. As a country develops the number of doctors increases because of the increase in levels of education.

Food and water security
As a country develops its food and water supply becomes more secure. This is because it has the technology to improve the intensity of agricultural production and to provide water supplies. It also has the wealth to buy in food if it cannot produce enough itself.

Education
Literacy levels relate directly to the level of development of a country. The lower the country's GDP, the lower the country's literacy rate.

The impact of uneven development on quality of life

Technology
As a country becomes more developed the level of technology in the country improves. One way of measuring this is by looking at the number of internet users in the country.

Employment
In developed countries the majority of the population work in tertiary and quaternary industry. In developing countries the majority of the population work in primary industry.

⬆ **Figure 10.11** The impact of uneven development.

	Literacy rate, 2015	Doctors per 1,000 people, 2010	Internet use per 100 people, 2013	Primary sector employment, 2012	Secondary sector employment, 2012	Tertiary and quaternary sector employment, 2012
Brazil	92	1.8	52	16	13	71
China	96	1.8	46	34	30	36
Namibia	82	0.4	14	16	23	61
Tanzania	68	0.8	4	70	5	25
Bhutan	65	0.1	30	56	22	22
USA	99	2.4	84	1	20	79
UK	99	2.7	90	1	15	84
Norway	99	4.2	95	2	20	78

⬆ **Figure 10.12** Social and economic data for selected countries.

⬆ **Figure 10.13** Access to housing.

Practise your skills

1 Select one developed country, one developing country and one emerging country from the table in Figure 10.12. Draw pie charts of their employment structure.
2 Justify your choice of developed country, developing country and emerging country based on just employment information.

Review

By the end of this section you should be able to:

✓ describe the impact of uneven development on the quality of life in different parts of the world: access to housing, health care, education, employment, technology, and food and water security.

ACTIVITIES

Study the data in Figure 10.12.

1 Which country has the highest number of doctors per person?
2 Which country has a high internet user rate compared to its literacy rate?
3 In your opinion, which of the housing environments in Figure 10.13 has access to a fresh water supply and sanitation? Give reasons for your answer.

Extension

1 Which country in Figure 10.12 has the best health care? Give a reason for your answer.
2 Study the information in Figure 10.12. Based on the information in this, section state two anomalies in the data. Justify your opinion.

Strategies that have been used to try to address uneven development

International strategies that attempt to reduce uneven development

International development aid has long been recognised as the way to help developing nations to grow out of poverty, but how should this aid be given? In 1970, the world's richest nations agreed to give 0.7 per cent of their **gross national income** (GNI) as official international development aid, annually. Since this time, wealthy nations have given millions of dollars each year, but few have achieved this target. Figure 10.14 shows the amount of aid given by countries in 2013 in US dollars and as a percentage of their GNI.

⬇ **Figure 10.14** Aid given in 2013.

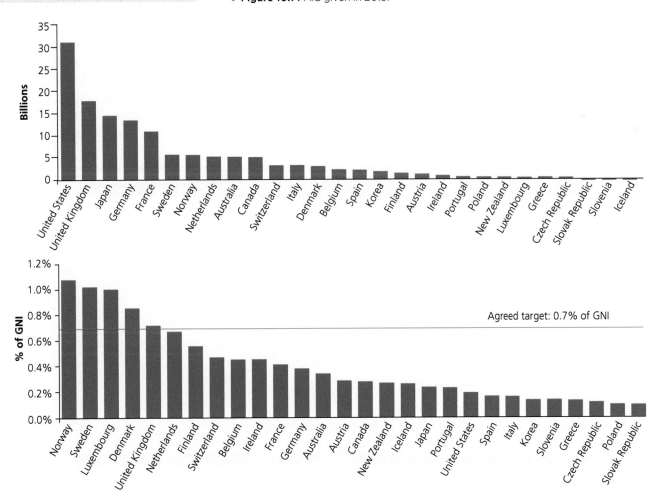

International aid

International aid can be given in a number of different ways.

Bilateral aid

The first aid that was made available to countries was after the Second World War. This was aid from USA to Europe to help rebuild its cities after the intense bombing that had taken place. This aid is known as bilateral aid and is given from one government to another government, usually with attached agreements, such as the recipient country has to give building projects to the donor country. An example of this is the money that India has loaned to Bhutan to build hydroelectric power schemes. India has provided the engineers and the technology and will get the electricity that is produced at a cheaper rate than the local people.

Multilateral aid

This is when developed countries give money to international organisations such as the **World Bank** or the United Nations. These organisations then redistribute the money in the form of loans to poorer countries.

Official and voluntary aid

Governments such as the UK and USA provide money that charity organisations can bid for to develop aid projects in different parts of the world. For example, the website in Figure 10.15 shows the UK government's international development funding that charities can apply for. One of the funds, UK

Aid Match, matches the money raised by charities pound for pound.

Voluntary aid

This is the money raised from donations and charities. Organisations such as Oxfam and Save the Children raise money through fund raising events, private donations and charity shops. This money usually funds **bottom-up projects**.

Inter-governmental agreements

These are agreements between developed world nations to work together to provide aid for developing countries. One of these agreements is with the EU. The EU delivers aid in different ways, one of which is sector support. As an example, the EU provides funds to develop education directly to the education department of the partner country, not to the government for them to use without direction from the EU. In other words, the EU will know what its money is being used to improve.

The development of emerging countries such as Brazil, China and India has seen these countries become major **donor countries** of aid. This is known as the South-South development co-operation, which has 30 countries that are donor members. They help develop countries that are not as well off as they are.

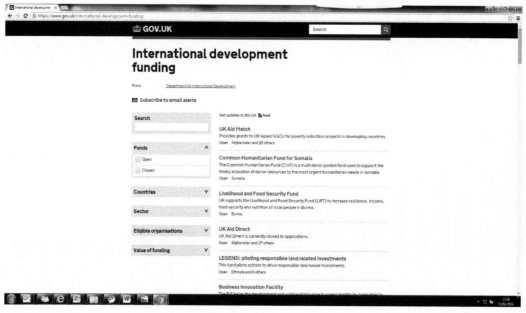

⬆ **Figure 10.15** UK international development funding.

The advantages and limitations of top-down and bottom-up development projects

Top-down schemes

This is development on a large scale: the government of a country will borrow money from organisations such as the World Bank in order to finance a large-scale scheme to benefit the whole country, such as the Three Gorges Dam in China. All decisions relating to the scheme are made by the government and the external groups involved. The local community have no say it what happens, although they are the people who will be most affected by the scheme.

Bottom-up projects

This is development on a small scale. They are schemes that are planned and controlled by local communities to help their local area. They are not expensive because they are smaller and use **appropriate technology**. Local people fund the schemes themselves or with help from aid groups.

Advantages	Disadvantages (limitations)
The country will develop more quickly because of the size of the projects.	The country will go into debt. In some cases these debts have never been paid off.
The scheme is run by the government so it is likely to achieve its development objectives.	The debt may have conditions attached to it which mean that the country is under external influences for many years.
In some cases, for example large HEP schemes, it is the only way to raise the capital due to the size of the project.	The end product is usually expensive to maintain.
It is a way of helping the large urban populations of a country, but often at the expense of the rural areas.	Much of the building work is done by machines or foreign companies so local jobs are not created.
	Local people have no say in what happens. In many cases they have lost land.

⬆ **Figure 10.16** The advantages and limitations of top-down projects.

China builds biggest dam in the world!

The Three Gorges Dam on the Yangtze River in Sandouping, China is the biggest hydroelectric power station ever built. At full capacity it can produce 22,500 megawatts to supply the large cities in the area. The dam has had a major impact on development in South West China. It has improved trade, allowing container ships further down the river; it controls flooding, so agricultural production has improved; and, of course, it provides clean, affordable electricity to thousands of people. There were limitations to the dam: 1.4 million people were moved from their ancestral homes. It cost the country US$22 billion and led to the extinction of the Yangtze river dolphin.

⬆ **Figure 10.17** Three Gorges Dam – a top-down project.

Advantages	Disadvantages (limitations)
The scheme is run by the local people so is likely to achieve its development objectives.	The country will develop more slowly because of the size of the project.
The end product is usually affordable to maintain.	It does not help the majority of the population who live in urban areas.
It is a way of helping the rural poor.	
Local people decide what happens to their community.	
Appropriate technology is used.	

⬅ **Figure 10.18** The advantages and limitations of bottom-up projects.

Micro-hydro projects help rural communities

Practical Aid is a charity that works with people in rural areas in developing countries to provide them with appropriate technology to improve their quality of life. One of their schemes is a micro-hydro project. This is a small-scale way of providing energy for rural communities from falling water such as steep mountain rivers. The scheme will generate up to 500 kilowatts of power and should last for over twenty years. The micro-hydro plants are owned and operated by the communities they serve with maintenance carried out by skilled members of the community. So, the scheme provides energy and employment. Any excess energy is stored in rechargeable batteries for villages that are further away from the scheme so they can also have power for workshop machines and domestic lighting. The scheme has little impact on the environment because there is no dam being built. It also cuts down the need for wood for fuel.

> Water flows down this channel to the forebay tank.

> Water from the river goes into a basin where the sediment settles on the bottom. Water for the hydro plant is taken from the top of the basin/tank.

> The water flows downhill from the forebay tank through a pipe called a penstock.

> The water flows into a power house where it turns a turbine to produce electricity.

⬆ **Figure 10.19** Micro-hydro scheme – a bottom-up project.

Review

By the end of this section you should be able to:

✓ describe the range of international strategies (international aid and inter-governmental agreements) that attempt to reduce uneven development
✓ know the difference between top-down and bottom-up development projects, and their advantages and limitations in the promotion of development.

ACTIVITIES

1 Give the names of three different charities that provide aid. State their mission statement.
2 What is bilateral aid?
3 Describe an example of voluntary aid. It could be something that you have done to raise money for an aid charity.
4 Explain the advantages and limitations of top-down development projects.

Extension

1 Why would governments give aid with certain conditions attached?
2 Discuss the pros and cons of top-down and bottom-up projects.

The level of development of Tanzania is influenced by its location and context in the world

LEARNING OBJECTIVE

To study how the level of development of Tanzania is influenced by its location and context in the world.

Learning outcomes

▶ To know the location and position of Tanzania in its region and globally.

▶ To understand the broad political, social, cultural and environmental context of Tanzania in its region and globally.

▶ To explain the unevenness of development within Tanzania and reasons why development does not take place at the same rate across all regions.

What is the location and position of Tanzania in its region and globally?

Tanzania is in eastern Africa. It has borders with Kenya and Uganda in the north; Rwanda, Burundi and the Democratic Republic of the Congo to the west; and Zambia, Malawi and Mozambique to the south. Figure 10.20 shows where Tanzania is in its region and globally.

The broad political, social, cultural and environmental context of Tanzania in its region and globally

Tanzania is the thirteenth largest country in Africa. It has 800 km of coastline and contains Africa's highest mountain, Kilimanjaro. It is mountainous and densely forested in the northeast. Central Tanzania is a large plateau with plains and arable lands while the eastern shore is hot and humid. Three of Africa's great lakes are partly in Tanzania: Lake Victoria, Lake Nyasa and Lake Tanganyika. It has sixteen national parks as well as many game and forest reserves; 38 per cent of its land area is set aside in conservation areas.

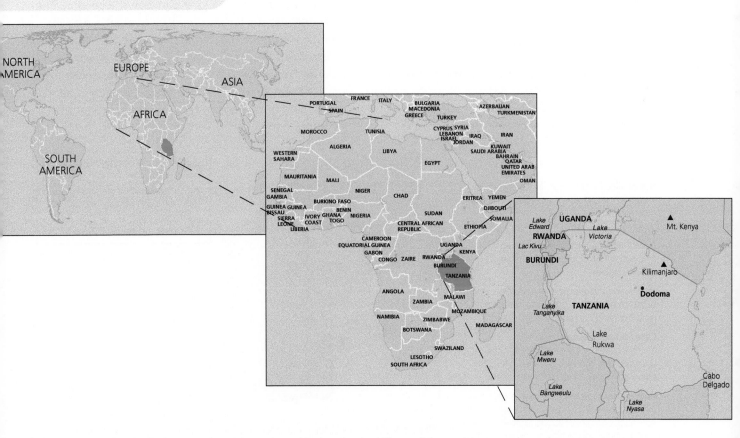

⬆ **Figure 10.20** The physical geography of Tanzania and its location in Africa and the world.

It has a population of 50 million and is one of the poorest countries in the world. Approximately 90 per cent of people living in poverty in Tanzania are based in rural areas. The population is unevenly distributed; most live on the northern border or eastern coast while the remainder of the country is sparsely populated. The country has two main languages: English and Swahili. English is used in schools because this is the language of the textbooks but Swahili is the official language of the country.

The former capital Dar es Salaam has most of the government offices. It is the largest city and port and is the country's wealthiest area. Dodoma, sited in the middle of the country, became Tanzania new capital in 1996 in an attempt to try to improve the standard of living in that area. On the United Nations' **Human Development Index** (HDI) in 2013 Tanzania was 152 out of 187 countries. It is estimated that one-third of the population live below the basic needs poverty line. The reduction in poverty is slow and Tanzania did not meet its target to halve it by 2015.

The country was a German colony from 1885 until the end of the First World War, when it came under British rule. It was then known as Tanganika with the Zanzibar Archipelago under separate colonial jurisdiction. When they became independent in the early 1960s, they joined to form the Republic of Tanzania.

In 1967 the government of the country realised that the country's wealth was leaking to other countries that had investments in Tanzania. It adopted a **socialist** political and economic approach, and **nationalised** all the banks and large industries. The government started to develop rural communities and subsidised services such as health and education. It formed an alliance with China to build infrastructure, including the 1,860 km TAZARA Railway from Dar es Salaam to Zambia. From 1980 the regime financed itself by borrowing from the International Monetary Fund and underwent some reforms. Since then the national economy has grown and poverty has been reduced. The first multi-party elections were held in 1995 when Benjamin Mkapa was elected president. The country returned to a **free market economy** and foreign investment was encouraged. Tanzania is now trying to develop through building infrastructure and continues to try to reduce poverty; it remains one of the poorest countries in the world.

Development within Tanzania is uneven: why does development take place at different rates in different regions?

The development of Tanzania has not been even across the country. This is shown in Figure 10.21, which shows the uneven GDP per capita in the country. The general reasons for uneven development are dealt with early in this chapter. In the case of Tanzania, the fastest rates of development have taken place around Dar es Salaam, which used to be the capital city and still is the main port. People can get employment in industries related to the port. In other parts of the country people rely on agriculture to earn their living.

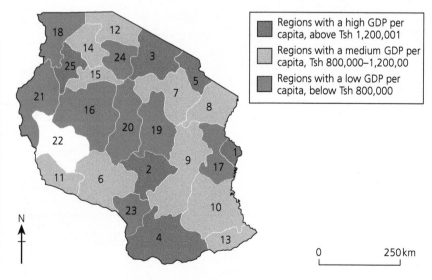

⬆ **Figure 10.21** A choropleth map showing the GDP per capita for the regions of Tanzania (Tanzanian shillings).

Number on map	Region	GDP per capita (in 1000s of Tsh)	HDI
1	Dar es Salaam	1,750	0.71
2	Iringa	1,425	0.69
3	Arusha	1,250	0.72
4	Ruvuma	1,210	0.67
5	Kilimanjaro	1,205	0.71
6	Mbeya	1,100	0.66
7	Manyara	1,000	0.63
8	Tanga	1,000	0.66
9	Morogora	990	0.62
10	Lindi	980	0.62
11	Rukwa	980	0.57
12	Mara	960	0.62
13	Mtwara	940	0.62
14	Mwanza	900	0.61
15	Shinyanga	820	0.54
16	Taboro	750	0.52
17	Pwani	740	0.54
18	Kagera	700	0.49
19	Dodoma	650	0.48
20	Singida	610	0.47
21	Kigoma	600	0.42
22	Katavi	No data	0.57
23	Njombe	600	0.69
24	Simiyu	610	0.55
25	Geita	1,500	0.55

⬆ **Figure 10.22** GDP per capita and HDI data for Tanzania's regions.

Review

By the end of this section you should be able to:

✓ describe the location and position of Tanzania in its region and globally
✓ understand the broad political, social, cultural and environmental context of Tanzania in its region and globally
✓ explain the unevenness of development within Tanzania and reasons why development does not take place at the same rate across all regions.

ACTIVITIES

1 Which is the official language of Tanzania?
  English Swahili German
2 How many countries border Tanzania?
3 Where is Africa's highest mountain?
4 Which is Tanzania's richest region?
5 On Figure 10.23, what is meant by the terms 'improved water' and 'adequate sanitation'?

Extension

1 Compare the state of development in different regions of Tanzania.
2 Why is the region closest to Dar es Salaam in the lowest group for GDP per capita whereas Dar es Salaam is in the highest group?

Practise your skills

Figure 10.21 is a choropleth map of Tanzania's GDP for each of its regions.

■ Describe the information shown in Figure 10.21.
■ Draw your own choropleth map for HDI for Tanzania's regions.

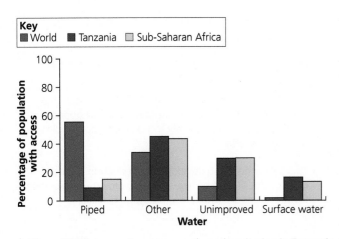

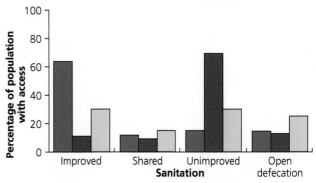

⬆ **Figure 10.23** Access to water supply and sanitation in Tanzania, Sub-Saharan Africa, 2010 and the world in 2012.

The interactions of economic, social and demographic processes influence the development of Tanzania

▢ LEARNING OBJECTIVE

To study the interactions of economic, social and demographic processes and how the influence the development of Tanzania.

Learning outcomes

▶ To know the positive and negative impacts of changes that have occurred in Tanzania's sectors of employment.
▶ To understand the characteristics of international trade and aid, and Tanzania's involvement in both.
▶ To explain the changing balance between public investment and private investment in Tanzania.
▶ To explain the changes in population structure and life expectancy that have occurred in the past 30 years in Tanzania.
▶ To know the changing social factors in Tanzania.

KEY TERMS

Transnational corporations (TNCs) – large companies that have their headquarters in one country and branches all over the world.

Public investment – money put into businesses by the government.

Private investment – money put into businesses by private financial backers.

The positive and negative impacts of changes to employment sectors on Tanzania's economy

There has been a growth in Tanzania's **informal sector** in both rural and urban areas by approximately 105 per cent between 2000 and 2006. This is not good for the economy as informal sector traders pay no taxes to the government.

Sector	Positive impacts	Negative impacts
Primary	Aid has been given to farming communities to try to introduce irrigation techniques which use appropriate technology. This should improve crop production to aid development. Mining of natural resources has brought much needed foreign investment into the country. The recent discovery of gas and oil will also help to provide much needed foreign investment in the future.	The agricultural sector's methods of production are out of date. For example, Tanzania uses an average of 9 kg of fertiliser per hectare whereas Malawi, at a similar stage of development, uses 27 kg and China used 279 kg. The sector is still dependent on the weather; dry years mean low crop yields. Agriculture as a share of GDP fell from 29% in 2001 to 24% in 2010. Improving productivity in this sector is crucial to the development of the country.
Secondary	This sector's share of GDP increased from 18% in 2001 to 22% in 2012. This will provide extra money for the economy.	Manufacturing's share of GDP is fairly constant with a slow growth rate. It is concentrated on a few goods which are of low value. Only 5% of employed people work in this area. This sector needs to develop the production of other goods through foreign investment to provide employment for those people who lack employability skills.

Sector	Positive impacts	Negative impacts
Tertiary	The service sector continues to grow with the development of a small middle class. Its growth rate was 8% between 2001 and 2012. With more people in higher paid jobs the country should receive more taxes to help it develop. There has been a growth in employment in education with the expansion of primary school education to all children. There has also been an increase in health care workers, providing employment for people who acquire the skills.	If the country is to develop, the service sector needs to continue to grow. Many of the jobs in the tertiary sector require a level of skill that will only be acquired through increased access to education.
Quaternary	The communications and financial services sector are the fastest growing in the economy, with a growth rate of 15% between 2003 and 2012.	This sector requires a highly skilled workforce, which required high wages. It does not provide employment for the low-skilled Tanzanian workforce. Its impact on the overall reduction of poverty is low.

⬆ **Figure 10.24** Tanzania's economy by sector.

The characteristics of international trade and aid, and Tanzania's involvement in both

International trade

In 2012 its top-three export trading partners were South Africa, Switzerland and China, with exports worth US$5.5 billion. Imports totalled US$11.7 billion. The main import trading partners were Switzerland, China and the United Arab Emirates. This meant the country had a negative **balance of trade** because its imports cost more than its exports.

The country has strong telecommunications and banking sectors, and its tourism industry continues to grow. It contributed 12.7 per cent of Tanzania's GDP and employed 11 per cent of the workforce in 2012. However, the economy is still based on agriculture, which makes up 24.5 per cent of GDP and 85 per cent of exports.

Industry and construction are growing, providing 22.2 per cent of GDP in 2013. This includes mining, manufacturing, electricity, gas and water supply. The country has started to extract diamonds, tanzanite

and gold to export. The coal extracted in Tanzania is used domestically.

International aid

Tanzania is the second-largest aid recipient in Sub-Saharan Africa, after Ethiopia. They received a combined total of US$26.85 billion in aid between 1990 and 2010. The main donors of aid to Tanzania are the USA and the UK, who gave US$793 and US$307 million dollars, respectively, in 2013. Other countries that give substantial amounts are Japan, Canada, Germany and Norway. Figure 10.27 on page 168 shows the main donors to Tanzania and the changes that have occurred since 2009.

Aid has been given to support the general budget of the country, but also to fund education, health care provision, water supply and sanitation (see Figure 10.28 on page 168).

KEY TERMS

Donor countries – countries that give aid.

Informal sector – people who set up informal businesses such as selling products on the street; they do not pay taxes or rent proper business premises.

Balance of trade – the difference between a country's imports and exports.

International Monetary Fund (IMF) – financial institution set up in 1945 to promote international trade. It pools funds from 188 member nations who can then make withdrawals when their economy is in difficulty.

World Bank – an international financial institution that provides loans to developing countries.

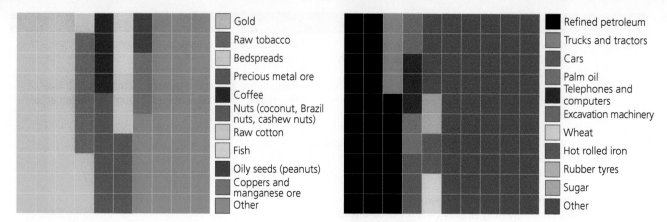

↑ Figure 10.25 Tanzania's major exports, 2012.

↑ Figure 10.26 Tanzania's major imports, 2012.

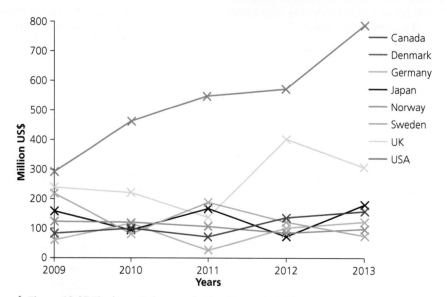

↑ Figure 10.27 The largest donors of aid to Tanzania, 2009–13.

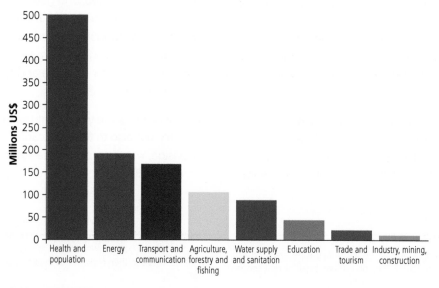

↑ Figure 10.28 The largest receivers of aid in the Tanzanian economy, 2013.

The changing balance between public and private investment in Tanzania

Over the past 30 years Tanzania has changed from a socialist to a market-driven economy. The government still owns some of the telecommunications, banking, energy and mining companies but it has encouraged international **private investment** in these areas. For example, international-owned commercial banks now own 48 per cent of the financial services in Tanzania. This has caused streamlining in the banking services and interest rates have come down. All land in Tanzania is owned by the government and people can rent it for 99 years. Overseas investors would like to buy land but the government is reluctant to sell. This is partly due to the natural resources that lie underground. Large **TNCs** are now responsible for the mining of **minerals** in the country such as gold, copper and silver. This is due to the government's law reforms, which made it easier to invest in the country. Domestic companies make up ten per cent of the mining industry. The recently discovered oil and gas fields are also being developed by international companies and are increasing the amount of private investment in the country.

In the past the **IMF**, World Bank and other donors have provided funds to improve the transport infrastructure (rail and ports) in the country. More recently, the government is trying to increase competition in the provision of infrastructure, such as electricity provision, to try and improve the service received. They are also trying to get private investment from domestic companies and TNCs in railways and water supply. Many of the regions of Tanzania have brochures on the internet to inform overseas investors about the great investment opportunities that are available in their regions.

Changes in population structure and life expectancy over the past 30 years in Tanzania

The total population of Tanzania was just over 51 million in 2015, see Figure 10.29. Approximately 70 per cent of the population still lives in rural areas. A large proportion of the population, approximately half, is under the age of 15. This is a large increase over the past 30 years due to the high birth rate, see Figure 10.31.

Life expectancy is improving but is still relatively low. It is very similar to countries at the same level of development. Malaria and diarrhoea are still some of the main reasons for death in the country among children, although the death rate is declining due to an improvement in health care financed by donor countries.

Year	Life expectancy (years)
1985	51
1990	50
1995	50
2000	50
2005	55
2010	59
2015	62

↑ **Figure 10.30** Life expectancy Tanzania, 1985 to 2015.

Practise your skills

Draw a line graph for the information given in Figure 10.30.

Current population	51,327,763
Male population (50%)	25,646,418
Female population (50%)	25,681,345
Births this year	1,005,881
Births today	1,906
Deaths this year	372,583
Deaths today	706
Net migration this year	−16,333
Net migration today	−31
Population growth this year	616,965
Population growth today	1,169

← **Figure 10.29** A population time clock for Tanzania, 24 August 2015.

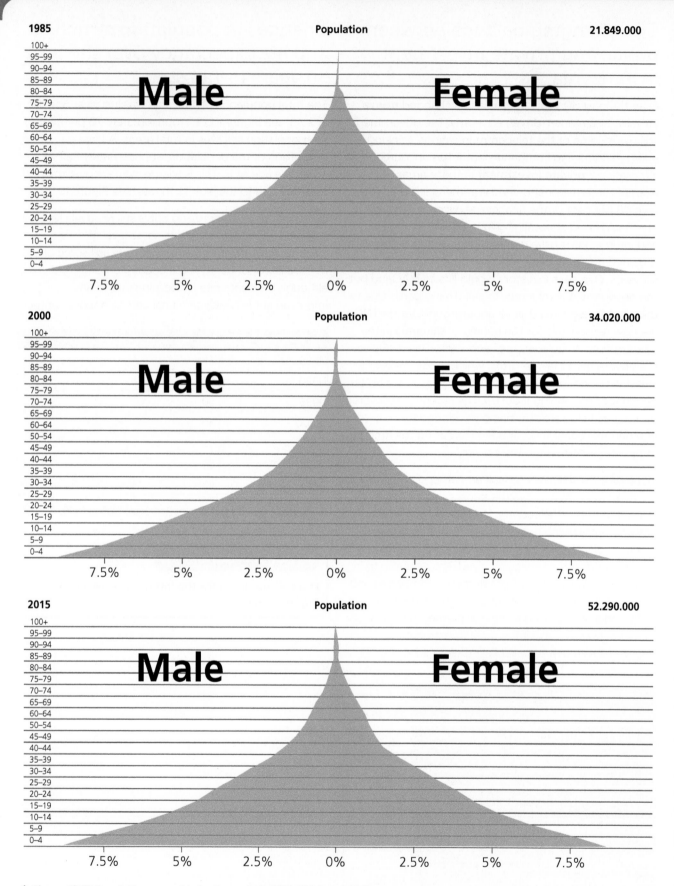

↑ **Figure 10.31** Population pyramids for Tanzania in 1985, 2000 and 2015.

Changing social factors in Tanzania

Tanzania has seen a growth in inequality which started to develop with the rich becoming richer and the poor living in greater poverty. However, the government is trying to address this problem and has achieved one of its goals: the provision of primary education for all.

Inequality and growing middle classes

The economy of Tanzania is growing but the growth is driven by an increase in productivity, not employment. Therefore, a large and increasing number of people are unemployed or underemployed. If the country is to improve the quality of life of the whole population, it needs to improve employment opportunities and to improve agricultural production. Many young people are moving to urban areas but cannot find work; they work in the informal sector instead, which means that they are still living below the poverty line.

In urban areas of Tanzania a small but growing middle class is developing. Approximately ten per cent of the population, this group is growing both in size and in its demands. They have a lot of political influence and are demanding things such as imported goods, cheaper electricity and better urban services and infrastructure. The government is trying to meet these demands because it does not want to these affluent people to leave the country; however, a strengthening middle class could lead to even greater inequality in Tanzanian society.

Education

Primary education in Tanzania is compulsory but attendance rates are approximately 80 per cent because it is very difficult to enforce school attendance in rural areas. Figure 10.32 shows the rise in literacy levels in the country since 1988.

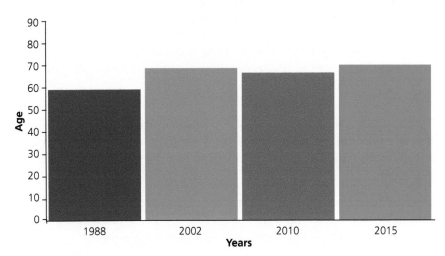

↑ **Figure 10.32** Literacy rate for Tanzania for selected years.

Review

By the end of this section you should be able to:

✓ describe the positive and negative impacts of changes that have occurred in the different sectors (primary, secondary, tertiary and quaternary) of Tanzania's economy
✓ understand the characteristics of international trade and aid and Tanzania's involvement in both
✓ explain the changing balance between public investment (by government) and private investment (by TNCs and smaller businesses) for Tanzania
✓ explain the changes in population structure and life expectancy that have occurred in the past 30 years in Tanzania
✓ know the changing social factors (increased inequality, growing middle class and improved education) in Tanzania.

ACTIVITIES

1 Study Figure 10.25.
   a) Which good is Tanzania's largest export?
   b) Which goods both supply four per cent of Tanzania's exports?
2 Who owns the land in Tanzania?
3 What is the informal sector of industry?
4 Suggest two reasons why children in rural areas do not attend school.

Extension

Between 1995 and 2000, life expectancy in Tanzania was stable. Since 2000 it has improved greatly. Suggest reasons for these differences.

The impact of changing geopolitics and technology on Tanzania

LEARNING OBJECTIVE

To study the impact of changing geopolitics and technology on Tanzania.

Learning outcomes

▶ To know how geopolitical relationships with other countries affect Tanzania's development: foreign policy, defence, military pacts, territorial disputes.

▶ To understand how technology and connectivity support development in different parts of Tanzania and its different groups of people.

KEY TERMS

Geopolitical – the influence of factors such as geography and economics on the politics and foreign policy of a state.

Refugee – a person who has been forced to leave their country in order to escape war, persecution or natural disaster.

The affect of geopolitical relationships on Tanzania's development

Foreign policy and military pacts

A country's foreign policy can be costly depending on the disputes it becomes involved in. Tanzania has never suffered from a civil war but it has been involved in other countries' disputes through foreign policy decisions that, on some occasions, have been costly. For example, the anti-Amin forces in Uganda in 1978 cost the country approximately US$500 million. In the past Tanzania has hosted **refugees** from neighbouring countries including Mozambique, Burundi and Rwanda, usually in partnership with the United Nations. This again can be a very costly endeavour if a country pays for the refugee camps itself; Tanzania had help from the United Nations.

Tanzania has always had good relationships with its neighbours. In recent years it has held peace talks to try to end the conflict in Burundi. It also supported the Lusaka Agreement to try to end the conflict in the Democratic Republic of the Congo. It has economic pacts with Uganda and Kenya, and is part of the Southern African Development Community.

Defence

The country has a small army, navy and air force. There are 25,000 regular personnel and 80,000 reserves. Their main work is being part of United Nations' peace-keeping missions to countries such as Lebanon and Sudan.

Tanzania has national service but it is only compulsory for people who want to work in government jobs or go to university. It lasts for anything up to two years. People can also volunteer to join the army for three years.

Territorial disputes

Tanzania has always been on good terms with its neighbours; however, there is an ongoing dispute with Malawi over the ownership of Lake Nyasa. The lake covers 30,000 sq km and, according to Malawi, was given to Malawi in the 1980 Heligoland–Zanzibar Treaty between Germany and the UK. Tanzania argues that the boundary between the two countries is in the middle of the lake. This means that Tanzania owns half of the lake and Malawi the other half. The dispute has been going on for years but was restarted in 2012 when Malawi gave a British company the right for oil exploration in the lake; the dispute has not been solved to date.

How does technology and connectivity support development in Tanzania and for different groups of people in the country?

The government of Tanzania has invested money to produce an ICT network for the whole country. The coverage of the network can be seen in Figure 10.33.

It will provide the necessary fibre cables for other network providers, such as mobile phone companies and broadband suppliers, to supply people in their homes. The network links to the cables come up from the sea bed in Dar es Salaam. It is hoped that the network will also link landlocked countries such as Uganda and Malawi to superfast fibre optic broadband.

Mobile usage in Tanzania has increased greatly over the last decade, with nearly 60 per cent of the population having mobile phones and many of them using the internet via their mobiles.

The use of the internet generally is lower, with about ten per cent of the population being connected, but the government sees connectivity as one of the main drivers of development and is ensuring that the infrastructure is in place.

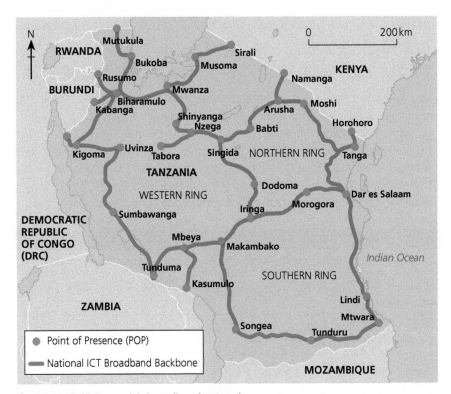

⬆ **Figure 10.33** Tanzania's broadband network.

Review

By the end of this section you should be able to:

✓ describe how geopolitical relationships with other countries have had an impact on Tanzania's development: foreign policy, defence, military pacts, territorial disputes
✓ understand how technology and connectivity support development in different parts of Tanzania and for different groups of people.

ACTIVITIES

1 Name one country with which Tanzania has had a border dispute.
2 Tanzania has helped refugees from a number of countries. Name two of these countries and research why the people took refuge in Tanzania.
3 What has been the impact of the introduction of the ICT backbone to Tanzania?

Extension

How is the government of Tanzania hoping to use connectivity to help develop the country?

There are positive and negative impacts of rapid development for the people and environment of Tanzania

■ LEARNING OBJECTIVE

To study the positive and negative impacts of rapid development for the people and environment of Tanzania.

Learning outcomes

▶ To know the positive and negative social, economic and environmental impacts of rapid development for Tanzania and its people.
▶ To understand how the government and people of Tanzania are managing the impacts of its rapid development to improve quality of life and its global status.

The positive and negative social, economic and environmental impacts of rapid development for Tanzania and its people

The development of Tanzania over the past 50 years has been steady but slow. Since the 1980s, however, great efforts have been made to improve the lives of poorer Tanzanians.

How are Tanzania's government and people managing the impacts of rapid development to improve quality of life and its global status?

Tanzania has for many years received vast amounts in aid from donor countries. It has used this money to provide primary education for all and to improve the country's infrastructure. However, even in urban areas 66 per cent of the population still lack access to piped drinking water, with the figure for proper sanitation being approximately 12 per cent. Some of these problems have been caused by rapid population growth but, if the country is to develop, the government must deal with these problems.

	Positive	Negative
Social	Improvement in life expectancy. Improvement in supplies of fresh drinking water and sanitation; 62% of the population now have supplied water and sanitation. All children have access to primary schools and attendance is above 80%.	Some rural areas are not benefited from improvements at all. In 2012, 28% of the population still lived below the poverty line. Because of the rapid expansion in schools teaching standards are low; 60% of students failed the secondary school leavers' exam in 2012. Health care is still poor with approximately 40% of the jobs not filled because of a lack of health care professionals in the country.
Economic	Improvements in GDP for the country. Foreign investment in the country is improving. Strong banking, financial and telecommunication sectors.	There is still inequality between regions. There are indications that there is still a large divide between rich and poor. In 2012 the richest 20% of the population accounted for 42% of total consumption, whereas the poorest 20% consumed only 7%.
Environmental	Electricity has been introduced to rural areas using bottom-up schemes; deforestation will slow down as less wood is used for fuel. Proper irrigation schemes using appropriate technology will allow the farmers to use their land more efficiently and stop overgrazing.	Gold mining causes problems of toxins leaking into water courses. Quarries are left as scars on the landscape. Deforestation due to rises in population numbers and the use of wood fuel for domestic purposes. Deforestation leads to loss of habitats and biodiversity. Overgrazing of farms in dry years is also a problem.

⬆ **Figure 10.34** Impacts of development on Tanzania.

The quality of life in Tanzania has not really improved for the majority of the population, who still live in rural areas and work in agriculture. The government is trying to help in these areas but development is slow and quality of life for the majority is not really improving. Over the next five years the government has pledged to put more money into helping the rural poor and improving the quality of life for people living in these areas by directing aid money into appropriate technology projects.

Tanzania's status in the global community is in some ways above that of its neighbours as it has never had a civil war and it has helped the UN on a number of occasions with refugees from countries that have internal problems.

On the corruption index on page 152, Tanzania is 119 out of the 174 countries listed – well below other countries at the same stage of development. However, big scandal broke out in the Energy Department in 2014. It was alleged that government officials were taking money and all aid was suspended for a number of months until the situation was sorted out.

Review

By the end of this section you should be able to:

✓ describe the positive and negative social, economic and environmental impacts of rapid development for Tanzania and its people
✓ understand how Tanzania's government and people are managing the impacts of its rapid development to improve quality of life and its global status.

ACTIVITIES

1 State one social and one economic negative impact of development in Tanzania.
2 Describe the positive and negative environmental impacts of development in Tanzania.
3 Explain how the government is trying to manage poverty in rural areas.

Extension

How successful has the Tanzanian government been in improving the quality of life for its population?

Examination-style questions

1 Define the term Human Development Index. (1 mark)
2 State **two** components of the HDI. (2 marks)
3 Figure 10.17 describes the Three Gorges Dam project. (2 marks)
   State which **two** statements below best describe the project.
   A The project is based on intermediate technology.
   B Local people are in charge of the project.
   C The government is in charge of the project.
   D The project has limited environment impact.
   E A large amount of money was borrowed to pay for the project.
4 a) Define what is meant by the term 'bottom-up project'. (2 marks)
   b) Explain **one** advantage and **one** disadvantage of this type of development. (4 marks)
5 Explain changes to life expectancy in Tanzania over the last 30 years. (4 marks)
6 Describe the changes shown on Figure 10.32. (3 marks)
7 Explain the negative environmental impacts of rapid development for Tanzania. (4 marks)
8 Assess the place of technology and connectivity on the development of Tanzania. (8 marks)
   Total: 30 marks

11 Resource Management

KEY TERMS

Biotic factors – the living organisms found in an area.

Abiotic factors – the physical, non-living environment, such as water, wind, oxygen.

Renewable energy – energy that comes from sources that can be reused or replenished and therefore will not run out.

Non-renewable energy – energy sources that, once they have been used, can never be used again.

Exploitation – the use of a resource in a non-sustainable way.

Overgrazing – rearing too many animals on land so that the roots of the vegetation are eaten. This means that the plants cannot regrow, leaving the soil bare.

A natural resource can be any feature or part of the environment that can be used to meet human needs

Natural resources can be defined and classified in different ways

Biotic factors are all the living things that are found in an area, whereas the abiotic factors are all the non-living things in an area, such water, wind and oxygen. If a natural resource is renewable it means that it can be used again and will not run out, for example the Sun or water. If a natural resource is non-renewable it means that it is finite in its supply. This means that there is a limit on the supply of the resource, for example coal.

The ways in which people exploit and change environments in order to obtain water, food and energy

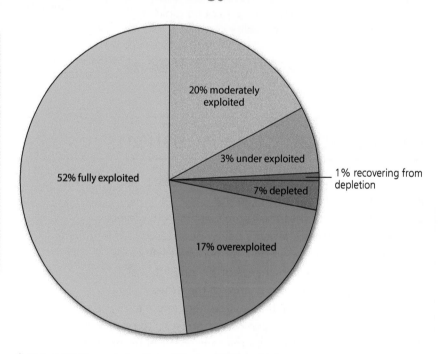

⬆ **Figure 11.1** General situation of the world's fish stocks.

	Exploitation	Changes
Water	Fresh water is an essential resource that is needed for people to survive. As the number of people in the world continues to increase, the need for water will also increase. Water is used for many things including drinking, washing and producing manufactured goods. In many countries it is not these uses that exploit water but the *misuse* of water sources, for example the extraction of minerals.	Some of the water we use comes from ground water sources. In many countries, including the UK, ground water is being used faster than it can be replenished by rain. This causes problems for plants and animals, and could cause a decrease in biodiversity in certain areas. When minerals are extracted, toxic by-products can be washed into rivers causing a decrease in the water quality of the water used for human consumption in the area. In many countries rivers are used for the disposal of sewage. In some cases the sewage is untreated before being put in the river.
Food	Farming – the land has been used to produce food for thousands of years; however, with increased population numbers in many countries, farming land is being overgrazed. Over the past 40 years forests in tropical countries have been cleared to make way for farming such as cattle ranching in the Amazon and palm oil production in Malaysia. Fishing – people take fish for food from the oceans. This has led to overfishing in many areas due to the demand being so great that the fish stocks cannot replenish themselves.	If land is overgrazed the bare soil is left exposed to the weather. The rain and wind can cause the soil to be eroded and either washed or blown away. Many tropical rainforests have suffered from deforestation. The trees have been felled to make way for farming. However, the land is only fertile for a few years. It is then left to be eroded by the heavy rains that fall each day. Overfishing has lead to a reduction of biodiversity in the oceans. As the ocean is a balanced ecosystem, if some fish species are reduced it has an impact on the whole ecosystem.
Energy	The extraction of fossil fuels to produce energy can cause a number of problems. Fossil fuels such as coal, oil and natural gas have been exploited for the energy they provide. The reserves of oil and natural gas have been dramatically reduced because of this exploitation, although there are still large reserves of coal.	The extraction and production of energy from fossil fuels can cause a reduction in air quality because of the gases that it produces, such as sulphur dioxide, carbon dioxide and carbon monoxide. Burning of coal to produce energy in the UK has caused acid rain to fall in Norway and Sweden. Trees in forests have died, resulting in a reduction in biodiversity.

⬆ **Figure 11.2** Changes caused by the exploitation of resources.

⬆ **Figure 11.3** Deforestation in Malaysia: forest being cleared for palm oil production.

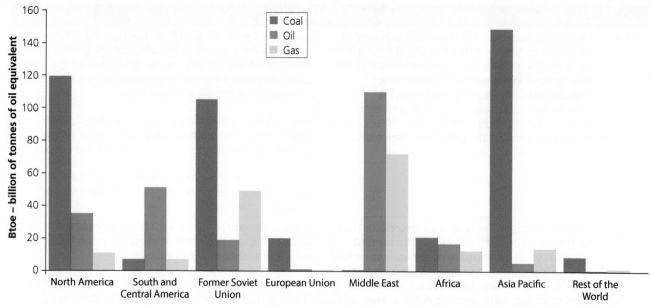

⬆ **Figure 11.4** Breakdown of fossil fuel reserves per geographical region, December 2012.

Type of fossil fuel	Percentage
Gas	20
Coal	51
Oil	29

⬆ **Figure 11.5** The world's reserves of fossil fuels, December 2012.

Practise your skills

Draw a pie for the information diagram in Figure 11.5.

Review

By the end of this section you should be able to:

✓ understand that natural resources can be defined and classified in different ways
✓ understand the ways in which people exploit environments in order to obtain water, food and energy
✓ know how environments are changed by this exploitation.

ACTIVITIES

1. Copy and complete the following table by adding two examples of each type of natural resource. Try not to use the ones mentioned in the text.

	Abiotic	Biotic	Renewable	Non-renewable
Example 1				
Example 2				

2. What is the difference between a renewable and a non-renewable resource?
3. Study Figure 11.1. What do you think is meant by the terms 'fully exploited', 'overexploited', 'moderately exploited' and 'depleted'?
4. What is the percentage of fish stocks that are fully and overexploited?
5. Describe how people exploiting environments for food can cause deforestation.

Extension

Explain how soil erosion has been caused by world population growth.

KEY TERMS

Fossil fuel – a naturally occurring fuel such as coal, oil and natural gas (methane) formed from the remains of dead organisms over millions of years.

Natural resources – materials that are provided by the Earth that people make into something that they can use.

Distribution – how something is shared out or spread across an area.

Arable farming – growing of cereal crops.

Pastoral farming – the rearing of sheep, cattle, pigs or any other animals on a farm.

Mixed farming – a farm which has cereal crops and animals.

Onshore oilfield – oil drilled on land.

Offshore oilfield – oil drilled from under the sea.

The patterns of distribution and consumption of natural resources vary on both a global and national scale

■ LEARNING OBJECTIVE

To study how the patterns of distribution and consumption of natural resources varies on both a global and national scale.

Learning outcomes

▶ To be able to describe the global variety and distribution of natural resources.

▶ To be able to describe the variety and distribution of natural resources in the UK.

▶ To understand global patterns of usage and consumption of food, energy and water.

The global variety and distribution of natural resources

The world provides a vast quantity of different resources which people use to provide food and other useful products. The **distribution** of these resources at a global scale can be displayed using maps; the UK will be dealt with later in this chapter.

Soil and agriculture

Soil regions of the world are shown on Figure 11.6. They are in very broad sweeping bands across countries and continents. The type of soil also relates to the climate and vegetation of the area. Some soils, such as chernozems and brown forest soils, are fertile and correspond with areas of high agriculture production, whereas other soils are less fertile. Agriculture in these areas is less productive.

Forestry

Forestry on a global scale is concentrated in certain areas. Figure 11.7 shows the countries which produce at least five per cent of the world's wood production. These countries include Canada, Brazil and the USA. Most other countries have their own forestry industry but it is too small to be incorporated on to the global map.

⬇ **Figure 11.6** Global soil regions.

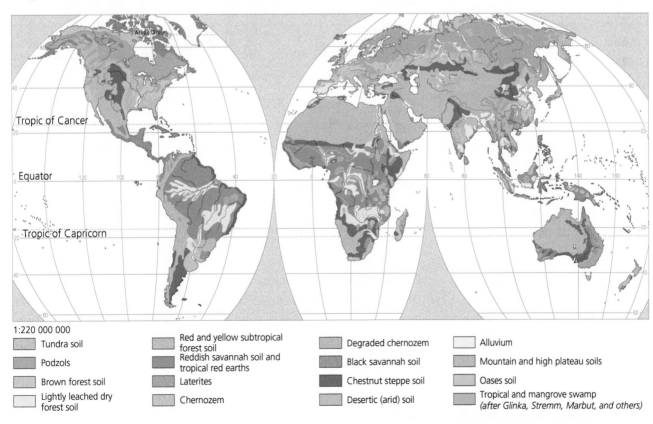

1:220 000 000

Tundra soil	Red and yellow subtropical forest soil
Podzols	Reddish savannah soil and tropical red earths
Brown forest soil	Laterites
Lightly leached dry forest soil	Chernozem
Degraded chernozem	Alluvium
Black savannah soil	Mountain and high plateau soils
Chestnut steppe soil	Oases soil
Desertic (arid) soil	Tropical and mangrove swamp *(after Glinka, Stremm, Marbut, and others)*

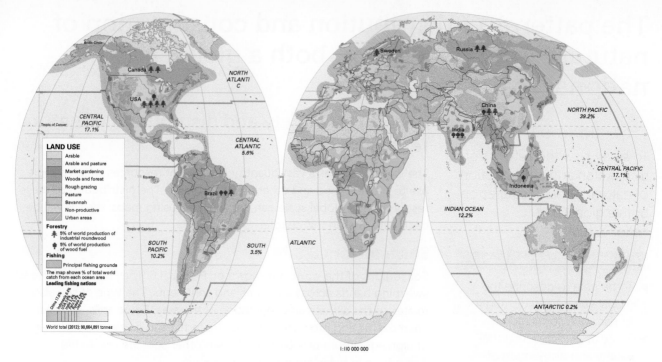

↑ Figure 11.7 Global land use and forestry.

Fossil fuels

Fossil fuel reserves for the world are shown on Figure 11.8. The countries that have the most oil reserves are Venezuela, Saudi Arabia and Canada. The countries with the highest gas reserves are Russia, Iran and Qatar. The USA has the highest reserves of coal left in the world although Russia and China also have vast reserves.

Rock and minerals

Rocks can be igneous, sedimentary or metamorphic and are distributed around the globe. These main categories can be broken down into hundreds of different types of rocks. How these rocks form is described in Chapter 1, pages 2–7. The most common rock type on the surface of the Earth is sedimentary. This layer of rock is very thin and goes about 2 km down into the crust. Below this level the most common rocks are igneous and metamorphic, although some sedimentary rocks occur here as well.

Minerals are distributed across the world, although some are concentrated on certain continents; for example, diamonds are found in Sub-Saharan Africa,

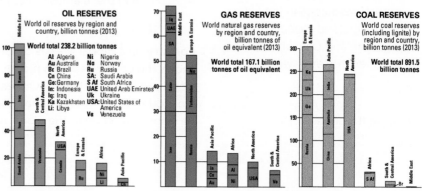

↑ Figure 11.8 Global fossil fuel reserves.

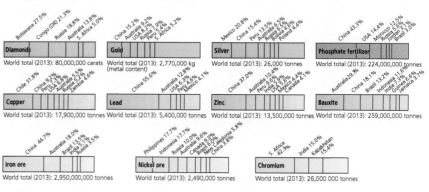

↑ Figure 11.9 Mineral production.

Russia and Australia but have not yet been discovered on other continents. Other minerals, such as iron ore, are distributed on all continents except Africa. The spread and availability of minerals has an impact on their price. Due to their rarity, diamonds are much more expensive than iron ore, which is much more abundant.

Water supply

The water that is used for human consumption is from rainfall, rivers, ground sources and, in some cases now, from the sea (see Chapter 13). However, the concern is that there will not be enough water for a global population expected to reach eight billion by 2025. In some countries expensive water transfer schemes have already been built because the available water supply is not where the majority of the people live. The availability of fresh water is becoming a major concern for many countries in the world.

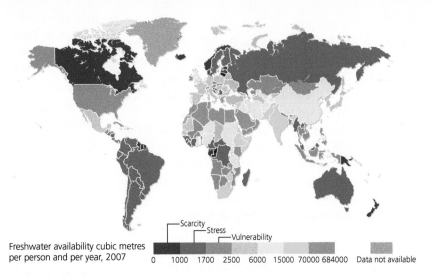

Freshwater availability cubic metres per person and per year, 2007

Scarcity — Stress — Vulnerability

| 0 | 1000 | 1700 | 2500 | 6000 | 15000 | 70000 | 684000 | Data not available |

⬆ **Figure 11.10** Global availability of fresh water, 2008.

The variety and distribution of natural resources in the UK

The UK has been using resources from the land and sea since early human occupation. As industrial processes developed, so did the technology to find uses for the different resources that were found within its shores.

Soil and agriculture

The UK has varied soils, many of which are very fertile. This means that farmers in the UK have a wide choice of crops or animals that they can farm. A map of the soil types can be seen in Figure 11.11. Many different kinds of **arable** and **pastoral farming** are practised, as can be seen in Figure 11.12. Some farmers have also started to grow vines and British wine is now being produced in Kent, Sussex and Devon. In Cornwall farmers have even begun to grow and make British tea.

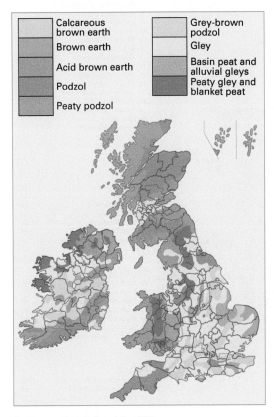

Calcareous brown earth
Brown earth
Acid brown earth
Podzol
Peaty podzol
Grey-brown podzol
Gley
Basin peat and alluvial gleys
Peaty gley and blanket peat

⬆ **Figure 11.11** Soils of the UK.

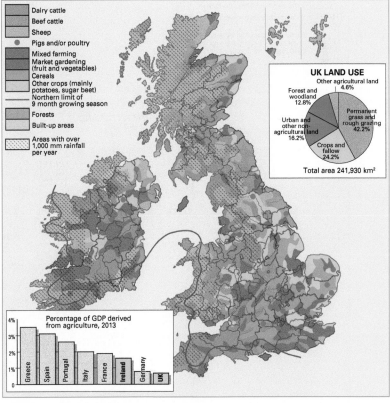

Dairy cattle
Beef cattle
Sheep
Pigs and/or poultry
Mixed farming
Market gardening (fruit and vegetables)
Cereals
Other crops (mainly potatoes, sugar beet)
Northern limit of 9 month growing season
Forests
Built-up areas
Areas with over 1,000 mm rainfall per year

UK LAND USE

Other agricultural land 4.6%
Forest and woodland 12.8%
Permanent grass and rough grazing 42.2%
Urban and other non-agricultural land 16.2%
Crops and fallow 24.2%

Total area 241,930 km²

Percentage of GDP derived from agriculture, 2013

Greece, Spain, Portugal, Italy, France, Ireland, Germany, UK

⬆ **Figure 11.12** Types of farms in the UK.

Forestry

Woodlands are distributed across the UK. Some of the woods are in private ownership and others are owned by the Forestry Commission. The Forestry Commission was set up because of the vast amount of deforestation that had occurred in the UK during the First World War. The country now has a forestry industry which employs approximately 800,000 people. It makes up approximately 2.5 per cent of the British economy. Forestry is concentrated more in the north and the west of the country where the land and climate is less agreeable and therefore more difficult to farm.

Fossil fuels

Fuels such as coal, oil and gas are all found in the UK. Coal has been mined in the country for hundreds of years and was the first source of fuel to power steam engines. The coal reserves in the UK were vast but most were a long way below ground. Coal was found in South Wales, Kent and the Midlands (Staffordshire, Yorkshire, Derbyshire, Nottinghamshire). Coal reserves were also mined in Northumberland and Durham, Scotland and Northern Ireland. Today the coal industry in the UK is very small and concentrates mainly on extracting coal from the surface, known as open-cast mining.

Oil and gas have more recently been discovered under the North Sea. The UK has one **onshore oilfield** at Wytch Farm on Poole Harbour.

Water supply

Although there is a plentiful supply of rainfall in the UK (see Chapter 13), it doesn't tend to fall in the places that have the highest concentration of population. This means that certain areas of the UK can become short of water in the summer when demand is high, especially if there has been a dry spring.

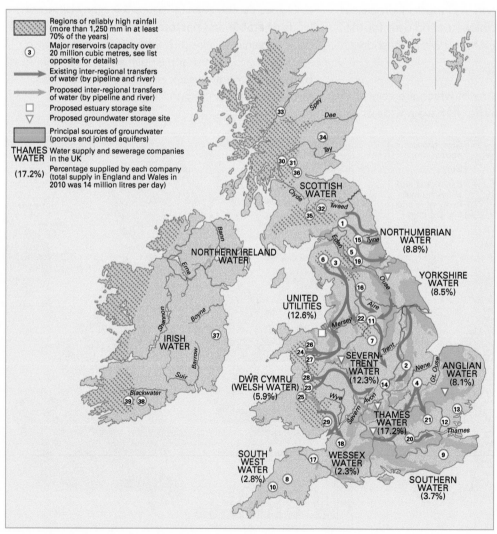

MAJOR RESERVOIRS
(with capacity in million m³)

England

1	Kielder Reservoir	198
2	Rutland Water	123
3	Haweswater	85
4	Grafham Water	59
5	Cow Green Reservoir	41
6	Thirlmere	41
7	Carsington Reservoir	36
8	Roadford Reservoir	35
9	Bewl Water Reservoir	31
10	Colliford Lake	29
11	Ladybower Reservoir	28
12	Hanningfield Reservoir	27
13	Abberton Reservoir	25
14	Draycote Water	23
15	Derwent Reservoir	22
16	Grimwith Reservoir	22
17	Wimbleball Lake	21
18	Chew Valley Lake	20
19	Balderhead Reservoir	20
20	Thames Valley (linked reservoirs)	
21	Lea Valley (linked reservoirs)	
22	Longendale (linked reservoirs)	

Wales

23	Elan Valley	99
24	Llyn Celyn	74
25	Llyn Brianne	62
26	Llyn Brenig	60
27	Llyn Vyrnwy	60
28	Llyn Clywedog	48
29	Llandegfedd Reservoir	22

Scotland

30	Loch Lomond	86
31	Loch Katrine	64
32	Megget Reservoir	64
33	Loch Ness	26
34	Blackwater Reservoir	25
35	Daer Reservoir	23
36	Carron Valley Reservoir	21

Ireland

37	Poulaphouca Reservoir	168
38	Inishcarra Reservoir	57
39	Carrigadrohid Reservoir	33

⬆ **Figure 11.13** Water supply in the UK.

Rock and minerals

The UK has varied rock types, which have already been discussed in Chapter 1. Figure 1.7a on page 5 is a map of the location of the main UK soil types. Other major rock types in the UK are clay and limestone.

The UK has a variety of minerals. They are used in construction to build houses and roads, as well as in industry, agriculture and horticulture. In 2013, 195 million tonnes of minerals were extracted from the UK landmass, which can be broken down into the following main categories:

- construction minerals 157 million tonnes
- industrial minerals 24.6 million tonnes
- fossil fuels 13.9 million tonnes.

Another 90.1 million tonnes of minerals were also extracted from under the sea (oil, gas, sand and gravel). The British Geological Survey website has an interactive map that gives information about the location of the mineral extraction industry (www.bgs.ac.uk/mineralsuk/maps/maps.html).

Understanding global patterns of usage and consumption of food, energy and water

Food consumption

Developed countries have the highest levels of food consumption, with much of Europe being in this category. Developing countries have the lowest levels of food consumption per person, for example some countries in Sub-Saharan Africa.

Practise your skills

Study Figure 11.14. You may need an atlas to help you answer these questions.

1 Which country in Sub-Saharan Africa has a food consumption per person of 3,000–3,500 calories a day?
2 Which country in Asia has a food consumption per person of under 2,000 calories a day?
3 a) What is the food consumption per day for the UK?
   b) Is there anything surprising about this figure?
4 Name a country in Asia with a food consumption of over 3,500 calories per person per day.

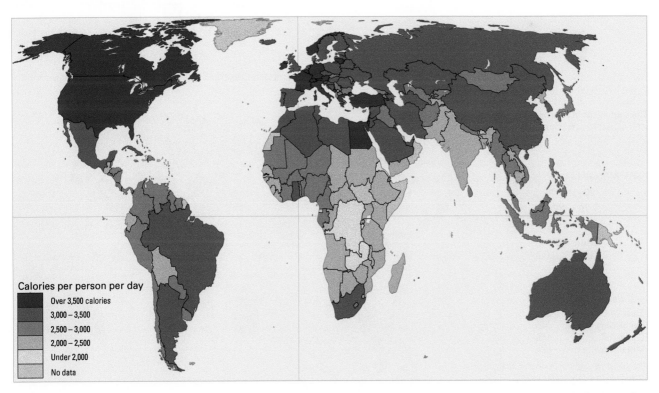

↑ Figure 11.14 Daily food consumption, 2014.

Energy usage

The amount of energy used by a country depends on many factors; one of these is the level of development of the country. Developed countries have a much higher demand for energy than developing countries; figures relating to this can be seen in Chapter 12. Emerging countries use large amounts of energy to power their developing industries. The demand for energy throughout the world continues to increase.

Water usage

The amount of water used varies greatly between different countries in the world. It also varies with the level of development of a country. The map in Figure 11.16 shows the amount of water that is available to be used in each country. The demand for water is dealt with in more detail in Chapter 13.

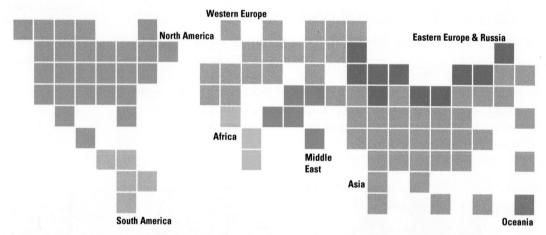

⬆ **Figure 11.15** Energy consumption per region. Each symbol represents 1% of world primary energy consumption.

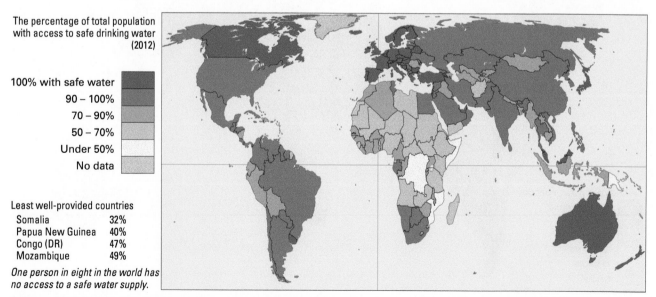

The percentage of total population with access to safe drinking water (2012)

100% with safe water
90 – 100%
70 – 90%
50 – 70%
Under 50%
No data

Least well-provided countries
Somalia	32%
Papua New Guinea	40%
Congo (DR)	47%
Mozambique	49%

One person in eight in the world has no access to a safe water supply.

⬆ **Figure 11.16** Global water supply.

Review

By the end of this section you should be able to:

✓ describe the global variety and distribution of natural resources
✓ describe the variety and distribution of natural resources in the UK
✓ understand global patterns of usage and consumption of food, energy and water.

ACTIVITIES

1 Name the countries on Figure 11.7 on page 180 that have at least five per cent of the world's production of industrial roundwood.

2 Study Figure 11.6 and Figure 11.7 on pages 179–80.

a) Copy and complete the following table by adding a type of farming for each of the soil types in the table.

	Podzol	Chestnut steppe soil	Chernozem	Mountain and high plateau
Type of land use				

b) Which area of the UK has the most arable farms? Give a reason for your answer.

Extension

Study Figure 11.12 on page 181. Explain the pattern of farming types shown on the map. (Hint: refer to climate and relief in your answer.)

Examination-style questions

1 a) Define the term 'biotic resource'? (1 mark)

b) Name **one** biotic resource. (1 mark)

2 State which of the following is not a fossil fuel.

A coal   B oil   C wind   D gas (1 mark)

3 Study Figure 11.15 on page 184. Calculate the difference between the percentage of energy consumed by North America and the percentage consumed by Africa.

A 20%   B 26%   C 3%   D 23% (1 mark)

4 Suggest **one** reason for this difference. (2 marks)

5 Suggest reasons why developed countries consume more energy than developing countries. (4 marks)

Total: 10 marks

Energy Resource Management

Renewable and non-renewable energy resources can be developed

Types of energy sources

There is an ever-increasing demand for energy in the world today. Energy is produced using a number of energy sources. These energy sources can be split into two different types: **non-renewable** and **renewable**. Non-renewable energy sources are ones that can never be used again once they have been used. As they took millions of years to form, they are known as finite resources; examples include coal and oil. Renewable energy sources are ones that can be reused and therefore will not run out; they are known as infinite resources and examples include the wind and the Sun.

What are the advantages and disadvantages of the production and development of coal, a non-renewable energy source?

Coal is used as an energy source all over the world. It is used to heat homes and is still the largest producer of electricity worldwide. The production and development of coal as an energy source has many advantages and disadvantages.

	Advantages	Disadvantages
Production	Coal is found in many countries around the world. It is cheap to mine. Much of the coal that is mined around the world is just below the surface and can be mined very easily (for example, Australia).	Waste heaps are left close to coal mines. Deep-shaft mining can be dangerous. Nine miners died in China in May 2012 when a shaft collapsed.
Development	It is relatively easy to convert it into energy by simply burning it. Coal supplies should last for another 250 years.	Acid rain is produced when coal is burnt. This has caused problems in the forests of Scandinavia. Greenhouse gases are emitted when coal is burnt.

⬆ **Figure 12.1** Open-cast coal mining in Australia.

⬆ **Figure 12.2** The advantages and disadvantages of the production and development of coal as an energy source.

What are the advantages and disadvantages of the production and development of wind power, a renewable energy source?

Wind power has recently been developed as a major source of electricity. The production and development of wind power has many advantages and disadvantages.

	Advantages	Disadvantages
Production	The wind is free. Turbines are relatively cheap, costing £1,500 for a 1 kilowatt wind turbine.	Some greenhouses gases are given off during the production of the turbine and when it is transported to its site.
Development	New wind turbines are quiet and efficient. It does not give off greenhouse gases. Wind turbines can be on land or at sea.	There needs to be an annual local wind speed of more than 6 m per second. They can be unsightly/visually intrusive. Offshore turbines may disturb the migration patterns of birds.

⬆ **Figure 12.3** The advantages and disadvantages of the production and development of wind as an energy source.

Review

By the end of this section you should be able to:

✓ classify energy resources as renewable and non-renewable
✓ describe the advantages and disadvantages of the production and development of one non-renewable energy source
✓ describe the advantages and disadvantages of the production and development of one renewable energy source.

ACTIVITIES

1 What is the difference between a renewable and a non-renewable resource?
2 Name one renewable resource.
3 Describe how wind has been developed as a resource.

Extension

Discuss the advantages and disadvantages of using renewable and non-renewable energy sources.

⬇ **Figure 12.4** Wind energy production in France.

Countries use energy resources in different proportions

LEARNING OBJECTIVE

To study how countries use energy resources in different proportions.

Learning outcomes

▶ To know the composition of the UK's energy mix.
▶ To understand that global variations in the energy mix are dependent on a number of factors.

KEY TERMS

Energy mix – the way that countries use energy in different proportions.

Energy consumption – how much energy is used.

Energy production – how much energy is being made.

Practise your skills

1 Draw a line graph of the information in Figure 12.6.
2 Describe the information shown in your graph. Use data in your answer.

Composition of the UK's energy mix

The UK's **energy mix** is made up of renewable and non-renewable energy sources. This mix changed quite significantly between 2007 and 2013 with the trend towards more renewables and coal-produced energy and a decline in natural gas and nuclear energy production.

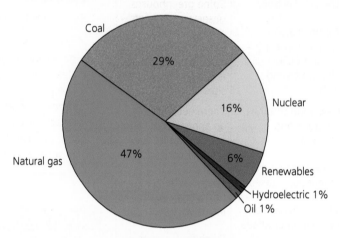

↑ **Figure 12.5** Energy sources in UK, 2010.

	1961	1971	1981	1991	2001	2011
Coal	80	56	73	66	35	31
Gas	0	1	1	2	37	45
Hydroelectric	3	2	2	2	1	1
Nuclear	2	11	14	21	23	16
Oil sources	15	30	10	9	2	1
Renewables	0	0	0	0	2	6

↑ **Figure 12.6** Changing UK energy mix, 1961 to 2011 (figures are percentages).

Global variations in the energy mix are dependent on a number of factors

Energy consumption varies greatly around the world, both in the amount consumed per person and in the way that energy is produced. This is due to a number of different factors including population, income and wealth, and the availability of energy supplies.

The global population increased by 27 per cent between 1990 and 2008, which means that a lot more energy was demanded. Added to this is the fact that the average use of energy per person globally increased by ten per cent. This means that there are more people who are also demanding more energy. This situation is not spread equally around the world:

countries with larger populations don't necessarily have the highest consumption of energy.

In Figure 12.7, Bangladesh and Brazil have similar populations but their energy consumption per capita is very different. This is due to other factors such as wealth and income. People in Brazil earn more money than people in Bangladesh, so are wealthier. They can afford to buy energy that is produced and the equipment that needs energy to make it work. In Bangladesh many do not have access to electricity or the income to buy electrical equipment. The USA has approximately twice the population of these two countries, but its energy consumption is many times more than these countries. This is mainly due to the wealth of the country, allowing it to provide the energy required by its population and the income of its population allowing it to buy the electrical equipment.

The availability of energy supplies is also important in the global energy mix. In Figure 12.7 Qatar has the highest total energy consumption; this is due to the availability of oil, which provides resources to produce energy but has also allowed the country to develop quickly and has provided wealth for its citizens. Not all countries with energy resources have a high per capita usage, however. Venezuela has rich oil fields but its consumption of energy per capita is fairly low.

Country	Total energy consumption per capita per annum (2003) {kgoe/a} kg of oil equivalent	Total population in 2005 (millions)
Bangladesh	161	141
Brazil	1,067	186.4
Poland	2,369	38.5
Qatar	21,395	1
Russia	4,423	143
Senegal	233	11.6
Singapore	5,158	4.3
USA	7,794	300
UK	3,918	60
Venezuela	2,057	26

⬆ **Figure 12.7** Total energy consumption per capita and population of selected countries.

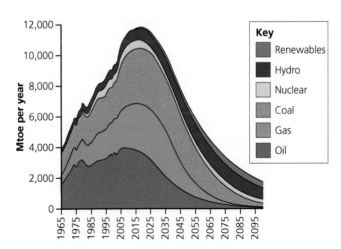

⬆ **Figure 12.8** The change in total global energy use for selected fuels from 1965 to predicted use in 2095.

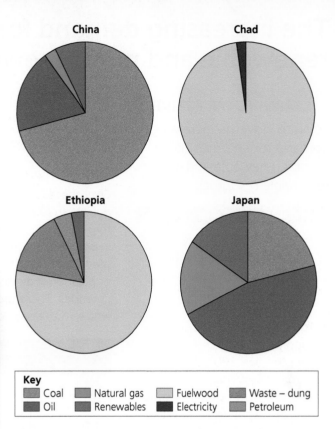

Key
- Coal
- Oil
- Natural gas
- Renewables
- Fuelwood
- Electricity
- Waste – dung
- Petroleum

⬆ **Figure 12.9** Energy mixes for China, Chad, Ethiopia and Japan.

Review

By the end of this section you should be able to:

✓ describe the composition of the UK's energy mix
✓ understand that global variations in the energy mix are dependent on a number of factors.

ACTIVITIES

1 Describe the UK's energy mix.
2 State two factors that affect the energy mix of a country.
3 Study Figure 12.7. Does the amount of energy consumed rise as population numbers in a country increase? Use information from the figure to explain your answer.

Extension

Explain why there are variations in the global energy mix between countries at different levels of development. Use information from Figure 12.9 in your answer.

The increasing demand for energy is being met by renewable and non-renewable resources

■ LEARNING OBJECTIVE

To study how the increasing demand for energy is being met by renewable and non-renewable resources.

Learning outcomes

► To know how and why global demand for energy has changed over the past 100 years due to human intervention.
► To know how and why global supply of energy has changed over the past 100 years due to human intervention.
► To know how non-renewable energy resources (coal, oil, natural gas and uranium) are being developed and how this can have both positive and negative impacts on people and the environment.
► To know how renewable energy resources (hydroelectric, wind and solar power) are being developed and how this can have both positive and negative impacts on people and the environment.
► To know how technology (fracking) can resolve energy resource shortages.

How and why has the global demand for energy has changed over the past 100 years due to human intervention?

Population growth

Over the last 100 years the population of the world has grown very quickly. In 1916 the global population was just under 2 billion, in 2011 it reached 7 billion, and by 2016 it should reach 7.4 billion. This acceleration in population numbers during the twentieth century has resulted in a great increase in the global demand for energy. The majority of the population growth has been in developing and, more especially, emerging countries. Therefore, demand for energy has increased in these countries which, in the past, did not have industry and had fewer energy demands.

Year	Population increase in billions
1801	1
1927	2
1960	3
1974	4
1987	5
1999	6
2011	7
2025	8

◄ **Figure 12.10** World population increase, 1801 to 2025.

Practise your skills

Study the data in Figure 12.10. What would be the best way to display this type of data? Give a reason for your answer.

Increased wealth

The population of the world is becoming increasingly wealthy, which has enabled people to afford technology that requires energy. For example, 100 years ago very few people had cars; people heated their homes with coal and only heated part of the house. Nowadays central heating uses energy and many families in developed countries own two cars. In emerging countries, people's living conditions are improving as they become wealthier, which has meant that they are using more energy.

Technological advances

Advances in technology during the past 100 years have led to an increase in demand for energy to power them. (see Figure 12.11)

⬆ **Figure 12.11** Inventions of the nineteenth and twentieth century.

How and why has the global supply of energy changed over the past 100 years due to human intervention?

Increased wealth and technological advances

The increased wealth in the world has allowed the development of new energy sources and has therefore increased the **supply** of energy. It has paid for the development of new technology to search out new reserves of energy to provide for the increase in **demand**.

The first major energy source was coal, which is still mined in many parts of the world. With the invention of the motor engine a new energy source was required – oil. As the private car revolution took off in developed countries, the demand for oil increased. New technological advances allowed this increase in supply: new reserves of oil and gas were discovered under the sea bed in places such as the North Sea and off the coast of Venezuela, South America. New technology was also developed to enable the reserves to be extracted.

New technological advances have continued to open up new types of energy sources, which have moved away from traditional fossil fuels. Energy sources such as wind, solar energy and the power of water both in rivers and the sea have all been harnessed as new energy sources due to advances in technology.

The development of non-renewable energy resources and its impact on people and the environment

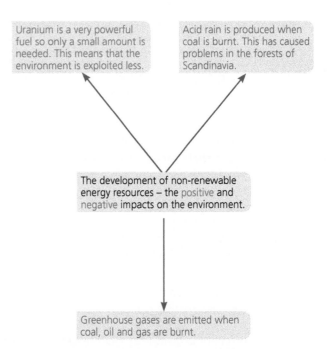

Many jobs are created when these resources are developed. For example, coal mines in South Wales in the nineteenth century, and the development of North Sea oil and gas in Scotland in the twentieth century.

The expansion of the nuclear power industry has provided many thousands of jobs.

Waste heaps are left close to coal mines. They can be unstable and collapse, for example Aberfan in 1966.

Deep-shaft mining can be dangerous. Nine miners died in China in May 2012 when a shaft collapsed.

Working with uranium can be very dangerous, for example the explosion at Chernobyl nuclear power plant in 1986.

The development of non-renewable energy resources – the positive and negative impacts on people

Coal is a very cheap form of energy for heating homes. Oil allows people to travel and access new areas, so it is a very useful energy source; it is also used as a way of heating homes.

Uranium is a very powerful fuel so only a small amount is needed. This means that the environment is exploited less.

Acid rain is produced when coal is burnt. This has caused problems in the forests of Scandinavia.

The development of non-renewable energy resources – the positive and negative impacts on the environment.

Greenhouse gases are emitted when coal, oil and gas are burnt.

⬆ **Figure 12.12** The impact of the development of non-renewable energy resources on people.

⬆ **Figure 12.13** The impact of the development of non-renewable energy resources on the environment.

The development of renewable energy resources and its impact on people and the environment

Fuel	Impacts on people (positive and negative)	Impacts on the environment (positive and negative)
Hydroelectric power	The reservoirs can be used for water sports or fishing. The reservoirs provide a water supply for areas that are located nearby. Often people have to move because their land is flooded to create the reservoir. The reservoirs are usually in remote areas, so people have to travel a long way to use the facilities.	It is less damaging to the environment as no greenhouse gases are produced. Large areas are flooded to create reservoirs to provide the water needed to drive the turbines. This will have an impact on the flora and fauna of the area.
Wind power	Home owners can have their own wind turbine so will benefit directly from 'free' energy. New wind turbines are quiet and efficient. They can be unsightly; many people see them as visually intrusive.	Wind turbines can be put out to sea so are less environmentally polluting. Wind turbines do not emit greenhouse gases once in use. They need a local wind speed of more than 6 m per second to be viable.
Solar power	They have no running costs so are a cheap source of energy. They can be fitted on homes so home owners have their own energy supply. They can be unsightly; many people see them as visually intrusive.	Can be fitted on to roofs so do not take up extra land space. Solar panels do not emit greenhouse gases once in use. If put in fields they are taking up valuable space where food could be grown.

⬆ **Figure 12.14** The impacts of renewable sources of energy on people and the environment.

How can technology (fracking) resolve energy resource shortages?

At its present rate of consumption, the UK will run out of its own supplies of oil in 4.5 years time. This makes the discovery of **fracking** – a way to access new supplies of gas and oil that were not in the original calculations from **shale rocks** – an important resolution to the problem of energy security in the UK. There are a number of issues around fracking however:

- It is a very water intense process, which may bring it into conflict with other water users, such as agriculture.
- The fuel is also very high in carbon compared to renewables and will add to the greenhouse effect. It equals coal in CO_2 emissions.
- A lot of energy is needed to obtain the fuel; the energy input to energy output ratio is a very low net gain.

The UK government is giving tax incentives to companies that develop fracking because they see it as a way of meeting their greenhouse gas emission targets by producing electricity from shale gas rather than coal.

Earth tremors shake Blackpool!

August 2013

Two earth tremors off the coast of Lancashire shook homes in Blackpool today. The tremors were at 5.37 a.m. (2.4 magnitude) and at 9.58 a.m. (3.2 magnitude). People wondered what was happening as their beds shook and pots rattled in their cupboards. No damages or injuries were reported.

The earth tremors have been blamed on the shale gas test drilling that is taking place. The company involved said 'it was related to the particular geology at one site' and that the tremors were 'unlikely to happen again'.

Protestors against shale gas drilling think otherwise and say it is just a sign of things to come.

Review

By the end of this section you should be able to:

✓ explain how and why the global demand for energy has changed over the past 100 years due to human intervention

✓ explain how and why the global supply of energy has changed over the past 100 years due to human intervention

✓ explain how non-renewable energy resources are being developed and how this can have both positive and negative impacts on people and the environment

✓ explain how renewable energy resources are being developed and how this can have both positive and negative impacts on people and the environment

✓ explain how new technology can resolve energy resource shortages.

ACTIVITIES

1   Create a table of impacts on people and the environment of the development of non-renewable resources. Add two more impacts of your own.

2   Find out more about one of the problems associated with the development of non-renewable resources, for example the disaster at Aberfan in 1966 or Chernobyl in 1986.

3   In groups of four, imagine that you are either a campaigner against wind turbines, a campaigner for wind turbines, a resident in the area where a wind farm is to be sited or the local planning officer. Devise a short play that highlights the points for or against wind farms.

Extension

'Fracking can solve non-renewable energy resource shortages.' Comment on this statement.

Meeting the demands for energy resources can involve interventions by different interest groups

☐ LEARNING OBJECTIVE

To study how meeting the demands for energy resources can involve interventions by different interest groups.

Learning outcomes

▶ To understand how attitudes to the exploitation and consumption of energy resources vary with different stakeholders.

KEY TERMS

Stakeholder – someone with an interest in what occurs.

Attitudes of different stakeholders to the exploitation and consumption of energy resources

⬇ **Figure 12.15** Differing views on the exploitation and consumption of energy resources.

We will raise a great deal of money in taxes if people continue to use oil, which will help to develop better hospitals and schools.

Government minister

There is plenty of oil left. We want to continue to extract oil because we provide lots of jobs.

Chief executive from the oil industry

We must build new homes with their own energy sources, such as solar panels, and plenty of insulation, so that each house consumes less energy.

Environmental campaigner

If I put solar panels on my land the government will pay me a grant. It is easier than farming and I make more money.

Farmer – landowner

Technology has got us out of messes before. Soon something will be invented to solve the problem of dwindling oil supplies. I've been reading in the papers about something called 'fracking'.

American Cadillac owner

Our country has developed using oil money. We must continue to exploit this resource if our country is to maintain its wealth.

Saudi Arabian spokesperson

Fracking for shale gas is not the solution to the UK energy crisis. We need energy based on renewable, not more fossil fuels which will add to greenhouse gas emissions.

Friends of the Earth campaigner

Review

By the end of this section you should be able to:

✓ understand how attitudes to the exploitation and consumption of energy resources vary with different stakeholders.

ACTIVITY

Choose one of the people in Figure 12.15. Research and prepare the argument for and against the statement you have chosen.

Management and sustainable use of energy resources are required at a range of spatial scales from local to international

Why do renewable and non-renewable energy resources require sustainable management?

If demands for energy in the future are to be met by the supply available, both renewable and non-renewable energy resources require **sustainable management**. This management must also take into account the effects on the environment of using some energy sources, such as global warming and the impact of waste. Countries also need secure energy supplies for geopolitical reasons.

'This looks like a good spot for non-renewable fuels'

⬆ **Figure 12.16** The global reliance on fossil fuels.

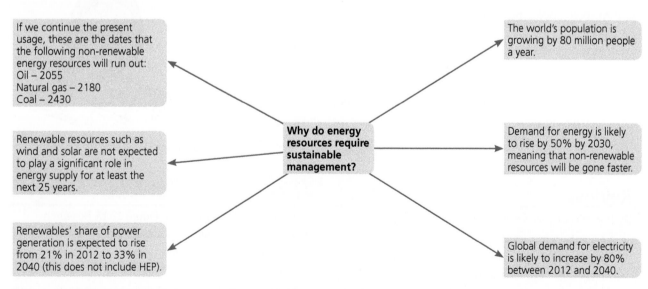

If we continue the present usage, these are the dates that the following non-renewable energy resources will run out:
Oil – 2055
Natural gas – 2180
Coal – 2430

Renewable resources such as wind and solar are not expected to play a significant role in energy supply for at least the next 25 years.

Renewables' share of power generation is expected to rise from 21% in 2012 to 33% in 2040 (this does not include HEP).

Why do energy resources require sustainable management?

The world's population is growing by 80 million people a year.

Demand for energy is likely to rise by 50% by 2030, meaning that non-renewable resources will be gone faster.

Global demand for electricity is likely to increase by 80% between 2012 and 2040.

⬆ **Figure 12.17** Why do energy resources require sustainable management?

What are the different views held by individuals, organisations and governments on the management and sustainable use of energy resources?

I have solar panels and an air-source heat pump to provide hot water and heat for my home. I try to manage the use of energy sustainably.

I do not want a wind farm on the hill opposite my house. I moved to Devon for the views and peace. They are noisy and unsightly.

Local resident

Home owner

We try to use hydroelectric power as much as possible to cope with peaks in demand because our coal-fired power stations take a long time to 'wake up'! This wastes a lot of energy.

We are upgrading our social housing with insulation and solar panels on the roof to help people to consume less energy.

Local government official

Energy company spokesperson

If all homes were constructed with energy efficiency in mind, less energy would be consumed, which means there will be more left for future generations.

We cannot keep using energy resources in the way that we are today because there will be none left for future generations. We need to be more sustainable in our approach and manage the resources we have left. We will provide incentives for people to use renewable energy sources and to insulate their homes.

Government minister for energy

Conservationist

⬆ **Figure 12.18** Differing views on the management and sustainable use of energy resources.

Located example How has Norway attempted to manage its energy resources in a sustainable way?

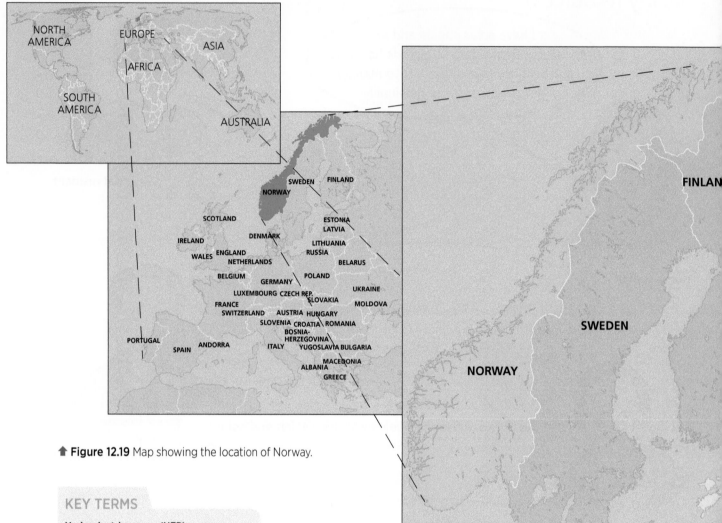

⬆ **Figure 12.19** Map showing the location of Norway.

Oil and gas production

Norway is a country in Northern Europe. It has vast reserves of oil and gas under the North Sea which it extracts and exports to other countries in Europe. It is the world's sixth-largest exporter of oil and second-largest exporter of gas. Norway invests a lot of money into research and development to improve extraction and transportation techniques of crude oil. This ensures that the extraction is managed in a sustainable way. For example, when these products are extracted from most oil fields up to 65 per cent is lost in the process; in Norwegian oil fields only 54 per cent is lost.

Alongside this it is reducing the amount of oil it produces: in 2001 it produced 3.4 million barrels; in 2013 it produced approximately half of this amount. The revenue it receives from the export of oil is being used to develop the renewable industry in the country. This is excellent management of the country's energy resources.

Renewable energy sources production

Norway is a mountainous country with reliable rainfall totals. For many years it has harnessed water to provide electricity – **hydroelectric power (HEP)**. At present 99 per cent of its electricity it produced from HEP. The government is not relying on this and has just started to harness wind power. They are investing money and allowing licences for eight new wind farms to be built onshore.

The management of Norway's energy consumption

Enova SF is an organisation that is run and funded by the government to promote energy savings in Norway. The Norwegian government has set a target to reduce greenhouse gas emissions by 30 per cent and to increase the renewable energy share of total energy consumption to 67.5 per cent by 2020. This can only be achieved by the whole country working together to sustainably manage their energy resources. The government is improving the national grid to make it a more efficient provider of electricity to the country.

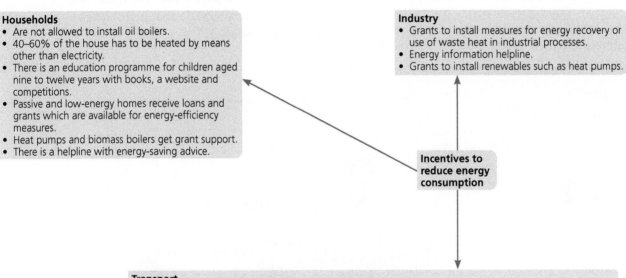

Households
- Are not allowed to install oil boilers.
- 40–60% of the house has to be heated by means other than electricity.
- There is an education programme for children aged nine to twelve years with books, a website and competitions.
- Passive and low-energy homes receive loans and grants which are available for energy-efficiency measures.
- Heat pumps and biomass boilers get grant support.
- There is a helpline with energy-saving advice.

Industry
- Grants to install measures for energy recovery or use of waste heat in industrial processes.
- Energy information helpline.
- Grants to install renewables such as heat pumps.

Incentives to reduce energy consumption

Transport
Since 2001 the government has been persuading people to buy electric cars.
There are a number of incentives:
- exception from the country's high taxes on fossil-fuel cars
- access to priority lanes reserved for taxis and buses
- free parking and charging on public parking places
- exemption from road tolls.

- Local initiatives in Norway include tolls on roads to stop people using them and speed limits. Car tax is also higher on cars that are less fuel efficient.
- Cities that improve their public transport systems get grants from the government.
- The country is investing money into the electrification of its rail network to move the country away from a reliance on fossil fuels.

⬆ **Figure 12.20** Incentives to reduce energy consumption in Norway.

⬆ **Figure 12.21** The Alta Dam, an HEP scheme in Norway.

Located example How has Bhutan attempted to manage its energy resources in a sustainable way?

Bhutan is a country in South Asia. It is landlocked, with China to the north and India to the west, south and east. It has a population of approximately 750,000. It is located on the southern slopes of the Eastern Himalayas with rainfall varying between 500 and 5,000 mm annually. This has allowed the country to develop HEP as an energy source.

The country is made up mainly of subsistence farmers and there is little industry. The government has started to encourage tourism but is restricting it to tourists who will not harm the environment or culture of the country. It has no reserves of oil or gas, so oil to power cars is imported. About 60 per cent of the population live in rural areas and use fuelwood as their main energy source. Over the last twenty years the government of Bhutan has attempted to manage its energy resources in a sustainable way.

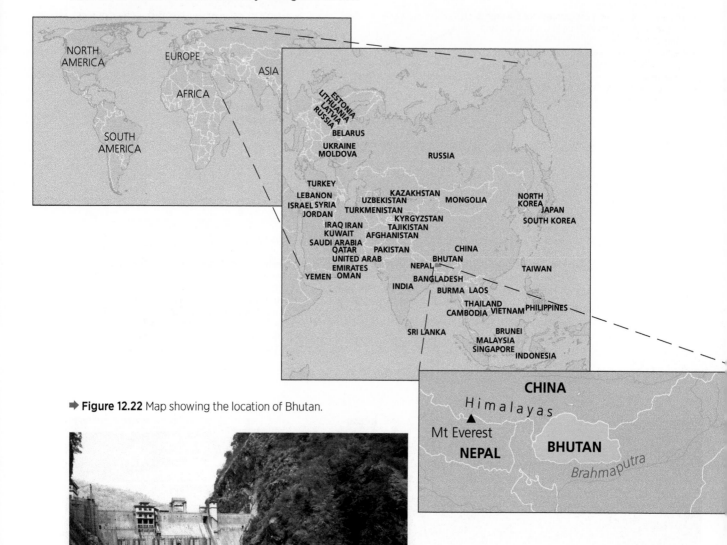

➡ **Figure 12.22** Map showing the location of Bhutan.

◀ **Figure 12.23** Tala HEP station in Bhutan.

HEP

Bhutan is developing its HEP resource as a means of revenue: at present it provides 40 per cent of the government's annual income. The Asian Development Bank has provided money to allow the government to build dams and construct the grids needed to both export the electricity but also to improve the electricity supply to rural areas in Bhutan. The excess electricity – about 70 per cent of what is produced – is exported to India. India has also helped Bhutan to develop its HEP potential by providing money, labour and expertise. This management of HEP production has allowed the country to raise capital to develop its infrastructure in a way that has not harmed its environment or culture. By 2013 Bhutan had provided electricity to 95 per cent of its households. Other countries, such as Austria and the Netherlands, have provided funds to extend the electricity grid to rural areas.

Other micro-energy schemes

The other five per cent of households live in remote rural areas. The government has decided to invest in small renewable energy schemes to bring electricity to these remote places to allow the whole of the country to benefit from electricity. The money has been provided by the Asian Development Bank who gave a grant to the Bhutan government. This was used to source 1,896 new solar home schemes in rural areas and to upgrade 2,500 that were already in existence. It also allowed for the development of two small wind turbines and a pilot scheme for biogas plants to supply 1,600 households.

⬆ **Figure 12.24** A micro-solar scheme in Bhutan.

Fuelwood

The majority of the population in both urban and rural areas use fuelwood as the main domestic source of energy, both for heating and to cook on. This is even with a supply of electricity, because fuelwood is freely available in the forests of Bhutan.

This is a problem in a number of ways both for the population and the environment. The people using it are subjected to damaging smoke inhalation, and it takes many hours to collect the wood and to produce enough heat to cook on open fires. It is also an environmental problem because of deforestation: the people do not replace the trees that they cut down. The burning of wood is also a global issue because it releases greenhouse gases. The government has tried to regulate the amount of fuelwood that people who do not have electricity are allowed to 16 m³ of fuelwood a year, and to 8 m³ a year if they do have electricity. Enforcing this is very difficult, however, because of the remoteness of the areas.

With the aid of outside agencies the government has started to address this problem and make the use of fuelwood as an energy source more sustainable. In 2013 the government and United Nations Development Programme started the Sustainable Rural Biomass Energy (SRBE) project. This promotes the use of biomass energy for cooking, heating and lighting in rural areas. Over 100 people from rural areas have been shown how to make brick-built stoves, which use less wood and burn more efficiently; they will then show others in their villages how to make the stoves. The project has a budget of over US$4 million and should benefit fourteen per cent of

rural population with 14,000 stoves being installed. It will reduce greenhouse gas emissions and reduce use of fuelwood through using more efficient cooking stoves. The stoves are smoke free so improve health; they also heat up more quickly so less wood is required which saves time and protects forests. There will also be a scheme to replenish 100 hectares of forest which has been cut down.

New policies

In 2013 the government issued its Alternative Renewable Energy Policy, which aims to promote the use of wind, solar, biomass and micro-hydropower systems. The use of wind energy will help if Bhutan suffers dry winters, as their HEP depends on rainfall and river flow rather than large storage reservoirs. This will stop the country being dependent on HEP in the future, making it more energy secure. This is good sustainable management of the country's energy resources.

The country is also working towards an Energy Conservation Policy to help households to conserve energy. This means that less energy will be required. One example of this is the SRBE project in rural areas. Another example is the introduction of electric cars. Nissan has come to an agreement with the prime minister of Bhutan to supply the government and a fleet of taxis with an electric car called the Nissan Leaf. Nissan will also provide charging stations across Thimphu, which is the capital of Bhutan. It is hoped that people will buy electric cars and therefore reduce Bhutan's reliance on imported oil and make use of the plentiful supply of electricity. This is making good use of a sustainable resource.

Energy source	Percentage of energy used
Fuelwood	42
Electricity	39
Diesel	8
Coal	5
Kerosene	2
Gasoline	2
LPG	1
Others	1

⬆ **Figure 12.25** The energy mix of Bhutan.

Practise your skills

Draw a pie chart of the information in Figure 12.25.

Consumption	Consumer numbers (households)
1995	21,800
1997	30,300
1999	33,700
2002	38,700
2004	46,100
2005	50,500
2006	61,300
2007	70,800
2008	77,200
2009	83,600
2011	91,700

⬆ **Figure 12.26** Electrical energy consumption by number of households.

ACTIVITIES

1 Copy out and complete the following table to compile a fact file for Norway and Bhutan.

Fact	Norway	Bhutan
Total population		
Land area		
Birth rate		
Death rate		
Life expectancy		
Urban population (%)		
Energy produced		
Energy consumed		
HDI		
Gross national income		
Internet users per 100 people		

2 Use this information to write a comparison of the state of development of the two countries.

3 Describe two ways that the Norwegian government is trying to reduce energy consumption.

Extension

Compare the way that Norway and Bhutan have managed their energy sources.

Review

By the end of this section you should be able to:

✓ explain why renewable and non-renewable energy resources require sustainable management

✓ understand the different views held by individuals, organisations and governments on the management and sustainable use of energy resources

✓ describe how Norway has attempted to manage its energy resources in a sustainable way

✓ describe how Bhutan has attempted to manage its energy resources in a sustainable way.

Examination-style questions

1 Identify the non-renewable energy source.
   A wind   B solar   C oil   D hydroelectric power   (1 mark)

2 Study Figure 12.5. State the percentage of the UK's energy mix that came from fossil fuels in 2010.
   A 93   B 77   C 48   D 29   (1 mark)

3 a) Study Figure 12.26. Calculate the difference in electricity consumers in Bhutan between 1995 and 2005.   (1 mark)
   b) Suggest reasons for the increase in consumer numbers in Bhutan shown in Figure 12.26.   (3 marks)

4 Explain **one** reason for the increase in energy consumption per person in the past 100 years.   (2 marks)

5 Explain the views of **two** different stakeholders, one for exploitation of energy resources and one against.   (4 marks)

6 Assess the impacts on the environment of developing non-renewable and renewable energy resources.   (12 marks)

Total: 24 marks

Water Resource Management

Fresh water supply varies globally

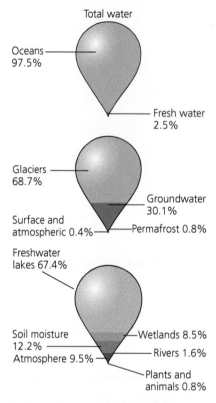

↑ Figure 13.1 The amount of water in the world.

The availability of fresh water varies on a global, national and local scale

The availability of fresh water is the amount of fresh water that can be easily accessed for human consumption, whether it is for domestic, industrial or agricultural use. One way of looking at this availability is to look at how the amount of rainfall varies between regions of the world. Figure 13.2 shows that rainfall totals vary globally and nationally. This will inevitably lead to variations in the amount of fresh water available. Variations on a local level are discussed on pages 210–11 when the availability of fresh water in the UK is looked at in more detail.

Another factor that restricts the amount of fresh water available is the fact that much of the water is away from the main concentrations of population. It has been estimated that the amount of fresh water available for human consumption is between 12,500 km³ and 14,000 km³ a year.

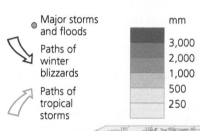

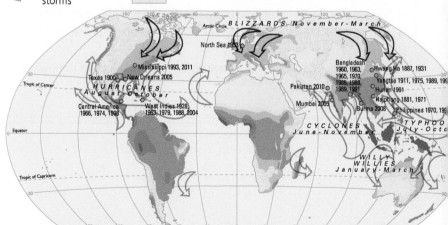

➡ Figure 13.2 Global annual precipitation.

Why do some parts of the world have a water surplus and others a water deficit?

A number of factors affect whether a region has too much or too little water. A region can have a physical surplus or deficit. This relates to the amount of rainfall that a region receives. Another factor is the evapotranspiration rate. Some regions with a reasonable amount of rainfall have very high evapotranspiration rates; this means that high temperatures will quickly turn the water back into a vapour, which rises back into the atmosphere. The water does not have time to enter water sources that make it available for human consumption.

A region can have an economic surplus or deficit. This relates to whether the government of an area can afford to provide water supply to the population. In developed countries the population has access to fresh water through the water supply system. This can lead to a water deficit if the available water is not used wisely. In developing and emerging countries the fresh water supply may be in surplus because the majority of the population does not have access to it due to economic water scarcity. Figure 13.3 shows the areas of the world that have a water surplus and those that have a water deficit.

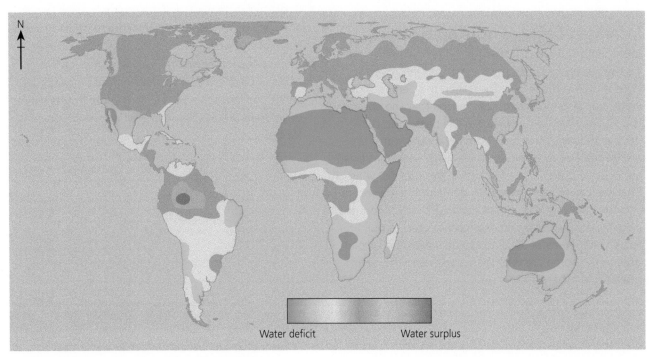

Water deficit   Water surplus

⬆ **Figure 13.3** Map of water surplus and water deficit.

How and why has the supply and demand for water changed in the past 50 years due to human intervention?

Water supply

Emerging or developing countries

The supply of piped fresh drinking water to households in many emerging or developing countries has increased in the past 50 years. This has been carried out by many charitable organisations, such as Water Aid UK, who have given people the means to provide themselves with drinking water as well as by organisations such as the World Health Organization. According to their estimates, 2.3 billion people gained access to improved drinking water between 1990 and 2012. But in 2012, eleven per cent of the global population still did not have access to clean drinking water. By 2015 the figure was down to 8 per cent.

Developed countries

The supply of fresh water to households in developed countries has changed little over the past 50 years due to human interventions. Although the variations in rainfall totals have had an impact on the amount of fresh water available in some developed countries.

Water demand

There has been an increase in the demand for water globally for a number of reasons:

- an increase in manufacturing industry in developing and emerging countries
- an increase in thermal electricity generation
- an increase in domestic use
- an increase in meat production; meat production uses eight to ten times more water than cereal production
- an increase in water for irrigation
- it takes an average 3,000 litres of water to produce one person's daily food intake, and the population of the world continues to grow.

Increased demand from emerging or developing countries

As the supply of piped fresh water has improved to households in these countries, then the demand for water has also increased. This means that the country is using its fresh water which in the past it did not have access to. This could possibly create water shortages. For example, in China fresh water has been supplied to many households over the past 50 years. This has caused an increase in demand in areas of the country that has less rainfall because this is where the population lives. The Chinese have started to build large water transfer schemes to deal with the surplus of water in some areas of the country and the deficit of water in others.

Increased demand from developed countries

As developed countries have become wealthier, the demand for water has increased. This is due a number of factors:

- Technological advances: dishwashers and washing machines use much more water than washing dishes and clothes by hand. The average dishwasher uses 3,000 litres of water a year.
- Changes to personal hygiene: 50 years ago houses did not all have bathrooms; many houses now have more than one bathroom. Many people now shower or bathe every day, which has caused a large increase in demand for water.
- Sport has increased: golf, for example, uses large amounts of water to keep the course green in some parts of the world.
- Leisure has increased: the building of swimming pools in people's gardens has increased, particularly in hot countries like Spain. Many more hotels have developed around the world, all with pools to cater for tourists.

Review

By the end of this section you should be able to:

✓ describe the global distribution of fresh water
✓ understand how the availability of fresh water varies on a global, national and local scale
✓ know why some parts of the world have a water surplus and other a water deficit
✓ know how and why the supply and demand for water has changed in the past 50 years due to human intervention.

ACTIVITIES

1. Describe the global distribution of water. Use Figure 13.1 on page 204 in your answer.
2. How much fresh water is there in the world?
3. How has the supply of fresh water changed in developing and emerging countries?
4. State three reasons why the demand for water has increased in developed countries over the past 50 years.

Extension

Compare the annual precipitation map in Figure 13.2 with that of water surplus and water deficit in Figure 13.3.

Differences between the water consumption patterns of developing countries and developed countries

The proportion of water used by agriculture, industry and domestically in developed countries and emerging or developing countries

Use	World (%)	Developed (%)	Developing (%)
Agriculture	70	30	82
Industry	22	59	10
Domestic	8	11	8

⬆ **Figure 13.4** World water usage.

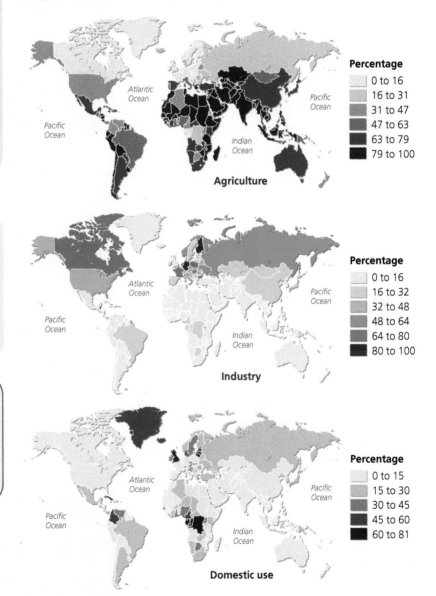

⬆ **Figure 13.5** Agriculture, domestic and industry water usage in 2000.

The amount of water used for agriculture (farming), domestic and industry varies greatly around the world. These variations are shown in Figure 13.5 on page 207. In more developed countries **water usage** for industry is higher than for agriculture, whereas in developing countries the amount of water used for agriculture is higher than for domestic uses.

Region	Water used daily per person (litres)
USA	400
UK	200
United Nations (recommended minimum)	50
Nigeria	40
Ethiopia	25

⬆ **Figure 13.6** Global daily water usage.

Why are there differences in water usage between developed and emerging or developing countries?

Agriculture

The amount of water used in agriculture is related to the type of farming being practised, the rainfall in the area and the state of development of the country. Developed countries where rainfall is low use **irrigation** systems, which require a lot of water. An automatic spray system can use 75 litres of water per second. They do not have the same technology in developing countries so the usage closer to 2 litres a second.

Industry

Industry's use of water in developed countries is on a large scale, with companies using millions of litres of water. For example, Walkers Crisps use 700 million litres of water a year at their factory in Leicester although 42% is recycled. In developing countries the industry is more small scale with most businesses run from home or small self build units. These **cottage industries** use much less water. However, due to large multinational companies moving their production to emerging and developing countries the percentage of water usage for industry in these areas will increase rapidly. It takes 3 litres of water to produce 1 litre of Coca-Cola. Coca-Cola have 24 manufacturing plants in India, a number of which have come into conflict with local people because of their over usage of ground water sources in the area.

⬆ **Figure 13.7** Developed world irrigation system.

⬆ **Figure 13.8** Developing world irrigation system.

⬆ **Figure 13.9** Walkers Crisps factory, Leicester.

KEY TERMS

Irrigation – the artificial watering of land for farming.

Cottage industry – small-scale production, often in a room of a person's home.

Domestic

Domestic water is used for many different purposes in developed countries (see Figure 13.10). These include indoor uses such as showering and laundry but also kitchen appliances and outside uses such as car washing and filling swimming pools. In developing and emerging countries many people still do not have piped water to their homes, although the numbers of people who do is increasing all the time. Many people spend many hours a day fetching water from wells and rivers. Their use of water is therefore restricted to what they can carry and its use is for cooking and personal hygiene.

15.7%

26.8%

1.7%

16.7%

Other 2.3%

21.7%

13.7%

1.4%

⬆ **Figure 13.10** Developed countries' domestic water usage.

Review

By the end of this section you should be able to:

✓ describe the proportion of water used by agriculture, industry and domestically in developed countries and emerging or developing countries

✓ understand why there are differences in water usage between developed and emerging or developing countries.

ACTIVITIES

1 What is the difference between the amount of water used for agriculture in the developing world and the amount used in the developed world?
2 Explain the different amounts of domestic water used in the USA and Ethiopia.
3 Keep a water diary for a day. Make a note of every time you use water and estimate the amount that you use.
4 Keep a note of what you eat in a day. Try to find out how much water is used to produce the things that you eat.

Extension

Explain why the irrigation systems used in the developed world uses more water than the irrigation systems used in the developing world.

Countries at different levels of development have water supply problems

■ LEARNING OBJECTIVE

To study how countries at different levels of development have water supply problems.

Learning outcomes

► To know why the UK has water supply problems.
► To understand why emerging or developing countries have water supply problems.

Why does the UK have water supply problems?

Imbalance of the supply from rainfall and the demand from population

The rainfall received by the UK is very varied. The north and west of the country receive the highest amounts, with Keswick in the Lake District receiving on average 1,500 mm of rain a year and London in the South East only receiving 550 mm a year. This means that the supply is plentiful in the north and west. However, one-third of the population of the UK lives in the South East, which is the driest part of the UK. The least populated areas of the UK are the mountain areas in Scotland and Wales. Here the annual rainfall totals are over 1,500 mm. This means that there is an imbalance between the areas with a plentiful supply and the areas with the greatest demand.

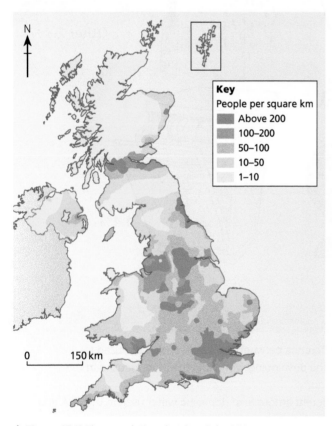

Key
People per square km
■ Above 200
■ 100–200
■ 50–100
■ 10–50
■ 1–10

⬆ **Figure 13.11** The population density of the UK.

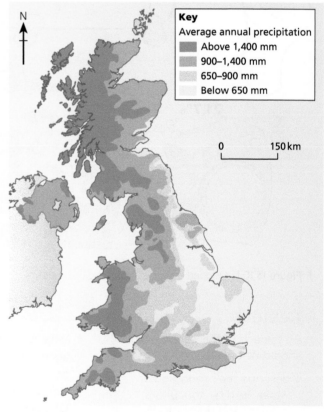

Key
Average annual precipitation
■ Above 1,400 mm
■ 900–1,400 mm
■ 650–900 mm
□ Below 650 mm

⬆ **Figure 13.12** UK average annual precipitation.

KEY TERMS

Imbalance – something that is not equal.

Population density – the number of people living in a certain area of land.

Infrastructure – in this instance it refers to pipes that bring water to our homes.

Ageing infrastructure: leakage to sewage and water pipes

Many of the pipes that supply water to households and industry in the UK are over 100 years old; some of them over 150 years old. The ageing pipes do not cause a problem with the quality of the water but there is a problem with leaks. In 2009, water firms in the UK lost 3.29 billion litres of water because of leaks: this is a third of the water that was supplied. Between 2004 and 2009, Thames Water reduced leaks by replacing old pipes by 27 per cent at a cost of £1 billion. In Cambridge, the water utility is replacing 13 km of water mains a year and installing meters on the pipes so that it can detect leaks more quickly. There will always be leaks on water pipes due to the environment and the pressure of road traffic, as most of our water mains go down roads, but the number of leaks must be reduced if the UK is to have enough water supply for the future.

The sewerage system of the UK also has an ageing **infrastructure** which, in many places, is over 100 years old. Before October 2011, much of the sewerage network was owned by the people who lived in the street, possibly without them knowing this! On 1 October 2011 most of the ownership of the sewage network passed to the water companies. They are now responsible for maintaining the network and mending leaks. Sewage leaks do occur, usually when old drains collapse due to heavy road traffic. This is because they were not built to withstand the volume and weight of today's traffic.

Seasonal imbalances

The UK receives most of its rainfall in the winter but the highest water demands are in the summer. The data in Figure 13.14 give average annual rainfall totals for the UK, although the exact amounts do vary with where you live in the UK, as explained earlier in this chapter. The monthly amounts show a similar pattern but there is definitely an increase in supply in the winter months of November–January. This will be when demand for water is at its lowest. The demand for water is at its highest in the summer months, especially if it is a hot summer. This can cause a problem of supply for the water companies, especially if the country has experienced a dry winter or spring, which has occurred in recent years.

⬆ **Figure 13.13** Collapsed sewer in Southport.

There can also be variations between yearly totals, as shown in Figure 13.15. In 2005 and 2006 there were long periods when the rainfall was below average. This causes problems because levels in water storage reservoirs go down. However, 2007 was an unusually wet year which brought the levels back up again.

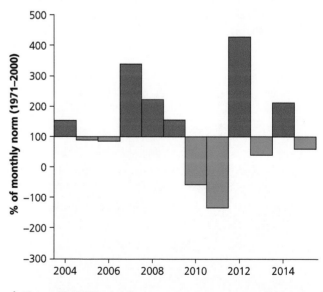

⬆ **Figure 13.15** UK rainfall per year as a percentage of the norm, 2004–15.

	Jan	Feb	Mar	Apr	May	Jun	Jul	Aug	Sep	Oct	Nov	Dec	Total
Rainfall (mm)	83	60	64	59	62	63	68	70	58	66	91	68	812

⬆ **Figure 13.14** UK average rainfall per month.

Why do emerging and developing countries have water supply problems?

Access to only untreated water

One of the main issues in the world today is the fact that many people still do not have access to clean water. This is due to the lack of piped water to households. The recent rapid development of China and India has improved access to water in these countries but many people in Africa still do not have a supply of clean, fresh water readily available.

- Between 1990 and 2012, 2.3 billion people gained access to improved drinking water.
- 748 million people still lack access to improved drinking water.

> In the Amazon region, waste materials from the mining and oil extraction industries is washed into the rivers, which has polluted the indigenous tribes' river water. The pollution can cause an increased risk of cancer, abortion, headaches and nausea. This is because their drinking water now contains toxins way above the level acceptable for human consumption.

> Another type of pollution is from untreated sewage. More than 80 per cent of sewage in developing countries is discharged untreated into rivers, lakes and coastal areas. The contaminated water can then cause many diseases.

⬆ **Figure 13.17** Sewage pipe putting pollution into a river near Mumbai, India.

Many diseases are related to drinking water that is not clean. The young and the old are particularly susceptible to diarrhoeal diseases, which are related to drinking dirty water, but these deaths, especially among children, decreased by nearly a million between 1990 and 2012.

Pollution of water courses

More than 840,000 people die each year from drinking water that has been polluted. Many people in emerging and developing countries still use rivers for their drinking water. The rivers are being polluted in many different ways.

⬆ **Figure 13.16** Water pit contaminated with waste from mining.

> Nearly 70 million people who live in Bangladesh have to drink water from wells. The water contains arsenic levels that are well above those which are recommended. This problem now affects 140 million people in 70 countries. The problem has been made worse because the ground water has not been allowed to replenish itself due to increased demand because of growing populations.

> Around 70 per cent of industrial waste in developing countries is disposed of untreated into rivers where it contaminates the water. In India and some African countries, many wells contain nitrate levels above the level recommended for human consumption due to intensive farming in the areas.

Low annual rainfall

Many developing and emerging countries are in parts of the world that have a low annual rainfall. This means that, as the population increases, these countries will have a physical scarcity of water. It is estimated that by 2025, 1.8 billion people will be living in countries with water scarcity. Many of these people will be in areas with low annual rainfall.

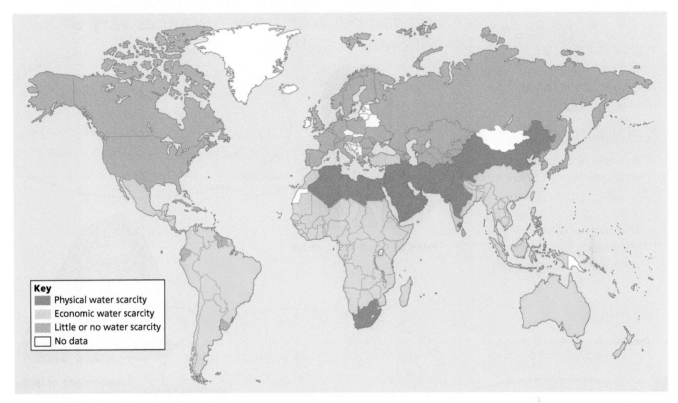

Key
- Physical water scarcity
- Economic water scarcity
- Little or no water scarcity
- No data

⬆ **Figure 13.18** Map of world water scarcity.

Review

By the end of this section you should be able to:

✓ explain why the UK has water supply problems
✓ explain why emerging or developing countries have water supply problems.

ACTIVITIES

1 Copy and complete the following table by naming water supply problems in different parts of the world.

Water supply problems in the UK	Water supply problems in developing or emerging countries

2 Choose one of the problems in the UK and explain why it is an issue.
3 Choose one of the problems in the developing world and explain why it is an issue.

Extension

Assess where the worst water supply problems occur, in developed or in developing countries?

Meeting the demands for water resources could involve technology and interventions by different interest groups

Attitudes to the exploitation and consumption of water resources vary with different stakeholders

If we are to keep our plants working efficiently and provide jobs for the local people, we need to use a lot of water. It is not our fault if this has to come from ground water sources; we have a business to run.

Head of Coca-Cola, India

The factories in our country will close if there are water shortages. We need to look after the welfare of our people and conserve water for domestic use.

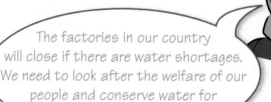

Government of India

Many countries in the world are short of water. We need to conserve water and use it sensibly if there is to be enough to go round.

We need to irrigate our crops so that we can make a profit and, of course, provide enough food for the people of America.

I need to dam the river if I am to create enough energy. It cannot be helped if this stops the river flooding, which used to provide irrigation water for the farmers downstream.

Water Aid campaigner

Farmer in Arizona, USA

HEP producer

⬆ **Figure 13.19** Different views on the exploitation and consumption of water resources.

If we keep using ground water supplies at the present rate there will soon be none left. We have to start to conserving our water supplies or there will be severe water shortages in dry years.

Mayor of London

They keep going on about water shortages. I cross the Thames every day on the way to work and there seems to be plenty of water in it. What about all the floods that happened last year?

London resident

Can technology resolve water resource shortages?

Many parts of the world have a shortage of water from rainfall but technological advances have come up with a possible solution. This is desalination, a process whereby salt is removed from sea water to make it drinkable. The concentration of salt in sea water does differ around the world, with the salty water costing more to make it drinkable, but it can be achieved.

There are currently 16,000 desalination plants worldwide producing roughly 70 million cubic metres of fresh drinking water per day. Saudi Arabia has the most desalination plants, with the USA in second place. The biggest problem with desalination is that it takes a lot of energy to desalinate a litre of sea water. However, the plants get around this by either using their own cheap supplies of oil and gas, as is the case in the Gulf states, by using cheaper night-time electricity or, more recently, by using solar power to operate the plants.

China has seven per cent of the world's fresh water but a fifth of its population. The country is looking towards desalination to provide for its people; Beijing, for example, aims to quadruple its sea water desalination capacity over the next ten years.

Thames Water built a desalination plant at Beckton on the Thames Estuary following droughts in the 1990s.

The plant can produce 150 million litres of water a day.

It takes water from the Thames.

The total cost of the scheme was £250 million.

The plant started to produce fresh water in 2010.

It is powered 100% by renewable energy.

It is not used everyday but it is used if there is a drought or levels in reservoirs or ground water supplies are low.

⬆ **Figure 13.20** The Beckton desalination plant.

The population of Saudi Arabia has quadrupled in 40 years and the government needs to supply its population with fresh water.

It produces just over 1 million m³ of fresh drinking water a day.

It takes water from the Persian Gulf.

It takes 2,400 megawatts of electricity a day to power the plant.

It started to produce fresh water in 2014.

It cost £3.5 million to build.

It is powered by solar energy during the day, which is when the peak supply is needed.

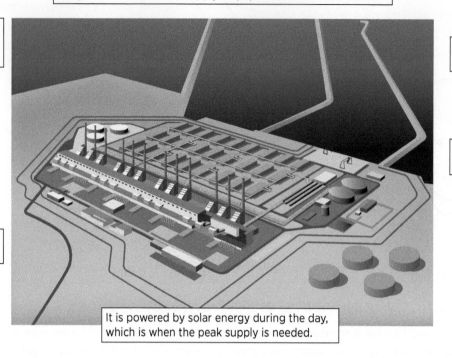

⬆ **Figure 13.21** A desalination plant on the Persian Gulf, Saudi Arabia.

Review

By the end of this section you should be able to:

✓ describe how attitudes to the exploitation and consumption of water resources vary with different stakeholders
✓ understand how technology can resolve water-resource shortages.

ACTIVITIES

Look at Figure 13.19 on page 214.

1 Is the Mayor of London for or against water conservation?
2 Draw a table with two columns headed 'water exploitation' and 'water consumption'. Complete the table using the views of stakeholders in Figure 13.19.
3 What is meant by the term desalination?

Extension

'Desalination can solve water resource shortages.' Comment on this statement.

Management and sustainable use of water resources are required at a range of spatial scales from local to international

LEARNING OBJECTIVE

To study how management and sustainable use of water resources are required at a range of spatial scales from local to international.

Learning outcomes

▶ To know why water resources require sustainable management.
▶ To know the different views held by individuals, organisations and governments on the management and sustainable use of water resources.
▶ To understand how the UK, a developed country, has attempted to manage its water resources in a sustainable way.
▶ To understand how China, a developing country, has attempted to manage its water resources in a sustainable way.

Why do water resources require sustainable management?

The global population continues to grow at a rate of 80 million people a year. These people need food and water. A person's daily use of water can be as little as 50 litres, but it takes 3,000 litres to produce the food for that person to eat. There is a finite amount of water in the world, therefore we must manage the water we have more carefully. The sustainable management of water resources will be different depending on the level of development of the country concerned.

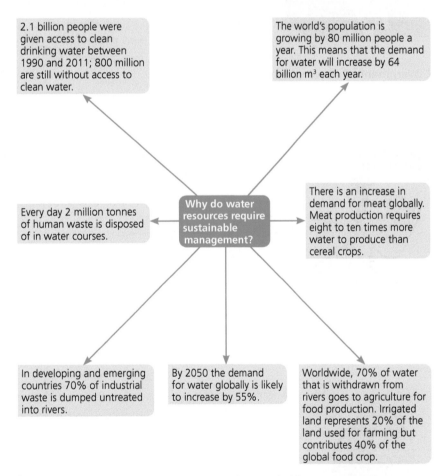

2.1 billion people were given access to clean drinking water between 1990 and 2011; 800 million are still without access to clean water.

The world's population is growing by 80 million people a year. This means that the demand for water will increase by 64 billion m³ each year.

Why do water resources require sustainable management?

Every day 2 million tonnes of human waste is disposed of in water courses.

There is an increase in demand for meat globally. Meat production requires eight to ten times more water to produce than cereal crops.

In developing and emerging countries 70% of industrial waste is dumped untreated into rivers.

By 2050 the demand for water globally is likely to increase by 55%.

Worldwide, 70% of water that is withdrawn from rivers goes to agriculture for food production. Irrigated land represents 20% of the land used for farming but contributes 40% of the global food crop.

↑ **Figure 13.22** Why do water resources require sustainable management?

The different views held by individuals, organisations and governments on the management and sustainable use of water resources

Government of developing country

> We need to provide fresh water for our people but will there be enough to go round?

> I need to use water to irrigate my crops or the harvest will fail and I will make no money. I am pleased that I have been shown new techniques to irrigate my fields which take less time and use less water.

Chinese farmer

UK resident

> Why shouldn't I use as much water as I like? It never stops raining!

> We need to work together to integrate water management so that we can get the most out of each river catchment area.

Conservationist, UK

African villager

> My life has been so much better since the government provided me with piped water.

> We are working as hard as we can to replace old Victorian water mains and to mend leaks as soon as they are reported to us.

Water company spokesperson

⬆ **Figure 13.23** Different views on the management and sustainable use of water resources.

Located example How has the UK attempted to manage its water resources in a sustainable way?

The water companies in the UK extract approximately 16 million m³ of water a day from water sources for domestic use. The demand for water has been increasing since the 1950s; the reasons for this increase are dealt with on pages 205–6.

Year	Unmetered households (litres per day)	Metered households (litres per day)	All households (litres per day)
2000	152	134	149
2002	157	137	150
2004	154	136	149
2006	153	134	147
2008	152	130	145
2010	158	129	144

⬆ **Figure 13.24** Domestic water usage.

Practise your skills

1 Draw a line graph for the information in Figure 13.24.
2 Why is a line graph the most appropriate way to display this data?

KEY TERMS

Aquifers – water-bearing rocks.

Metered households – meters measure the amount of water that a household uses and the home owner pays for the exact amount of water consumed.

Unmetered households – the water company estimates how much water the householder will use and charges for this amount.

The UK has a plentiful supply of water due to the amount of rainfall it receives. Earlier in this chapter (see page 210) we referred to the fact that the areas of the UK that receive the most rainfall are the north and west, but the areas with the highest population are the south and east. This is the challenge for the UK government and is one of the reasons why water resources need to be managed in a sustainable way.

For many years water has been transferred around the UK from areas that receive plenty of rainfall to areas with high populations. For example, the Elan Valley has supplied water for Birmingham since the early 1900s. Further and larger water transfer schemes would cost a lot to build and pumping water around the country would be very expensive. Water is also obtained from aquifers in the ground, which are water-bearing rocks. Over the past 30 years this supply has not been replenished as fast as it is being used by water companies. This is causing problems such as rivers running dry and ecosystems suffering.

There is also the uncertainty of climate change, which could bring more rainfall or could mean more droughts. As nothing is certain, decisions have been taken to encourage changes to the way that we view water.

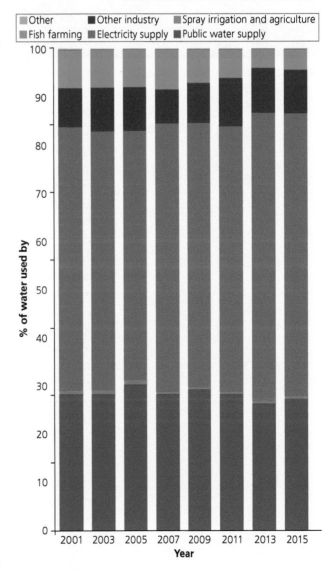

➡ **Figure 13.25** The uses of fresh water in the UK.

Solutions to the water shortage problem

The government has initiated a number of policies to attempt to manage the UK's water resources sustainably to ensure that there is enough for future generations.

⬆ **Figure 13.26** The Elan Valley reservoirs.

- Government has passed policies that ensure that water companies manage water sustainably. They have to produce 25-year plans which show their sustainable management of water sources.
- The government has set efficiency targets for water supply companies; their progress will be monitored each year by the government. Each company will have to develop plans to meet these targets.
- Many websites are available to help people to save water in their homes, some sponsored by the government.
- The government has begun to lower the licences granted to water companies for water extraction to allow ground water supplies to be replenished.
- A new scheme is to involve more people in the management of river catchment areas. This will focus on the management of land and water extraction in a sustainable way. Pilot schemes were set up by Defra in 2011 in certain areas of the country. The Environment Agency, the Rivers Trust, the Wildlife Trust and the water industry will come together to formulate plans for the management of the catchment area to give all stakeholders a say in the management of the water resource.

Since the government set targets on the water companies in the 1990s:

- leakage is down by 35 per cent
- between 2005 and 2010 water and sewerage companies in England and Wales laid, renewed or relined approximately 20,000 km of water mains.

Located example How has China attempted to manage its water resources in a sustainable way?

China has a serious water shortage problem. Over 400 Chinese cities are facing water shortages, with 136 experiencing severe shortages of water.

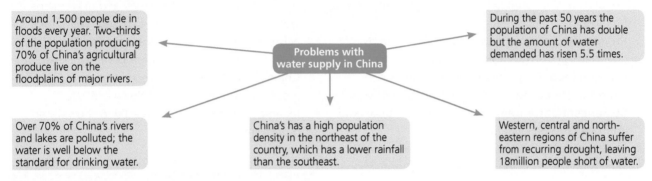

Around 1,500 people die in floods every year. Two-thirds of the population producing 70% of China's agricultural produce live on the floodplains of major rivers.

Problems with water supply in China

During the past 50 years the population of China has double but the amount of water demanded has risen 5.5 times.

Over 70% of China's rivers and lakes are polluted; the water is well below the standard for drinking water.

China's has a high population density in the northeast of the country, which has a lower rainfall than the southeast.

Western, central and north-eastern regions of China suffer from recurring drought, leaving 18million people short of water.

⬆ **Figure 13.27** Problems with water supply in China.

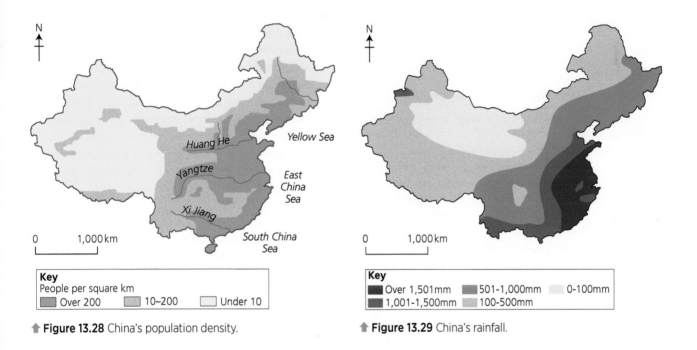

Key
People per square km
- Over 200
- 10–200
- Under 10

⬆ **Figure 13.28** China's population density.

Key
- Over 1,501mm
- 1,001-1,500mm
- 501-1,000mm
- 100-500mm
- 0-100mm

⬆ **Figure 13.29** China's rainfall.

Solutions to the water shortage problem

The government of China has initiated a number of policies to manage the water resources available sustainably to ensure that there is enough for their population and for future generations.

- Between 2010 and 2012 the Chinese government carried out a water census. It felt that without knowledge of the water situation it could not plan for the future. Around 800,000 surveyors were trained and sent around the country to find out how much water was being used in rural and urban areas. They also reported on the condition of lakes and rivers, and on any water conservation projects that were occurring.

- Desalination plants are being built on China's coastline to provide water for cities in the north and east at a cost of £2.1 billion. The desalination plants will triple the amount of water available for human use by 2020.

- Water is being redirected from China's wetter south to the north of the country which has a physical scarcity of water. The first section of the South-to-North Water Diversion Project opened in 2015 delivering water along the Beijing–Hangzhou Grand Canal from the Yangtze River in Jiangsu province to Shandong.
- Projects are being carried out in rural China by charities to cut the amount of water used to produce food. They work with local farmers to introduce improved irrigation methods that will save 65 per cent of total annual water consumption and improve crop production.
- The government has introduced a water-saving campaign using Olympic athletes to go into schools to teach the children the importance of saving water. They went into 1,000 primary schools in Beijing giving out leaflets and encouraged the children to post water-saving tips around the school.

- Beijing has a water conservancy museum which aims to show people how much water they use in their daily life. The hope is that it will heighten people's awareness of water and they will use less.
- In Shanghai 50 wells have been dug 240 m deep beneath large residential areas and universities. The water from these wells will be used when there are acute water shortages in the city. The city is also building new water treatment plants and reservoirs.
- China has also spent money on improving its reservoirs. By the end of 2015 over 50,000 reservoirs had been reinforced and their water quality improved.

Review

By the end of this section you should be able to:

✓ explain why water resources require sustainable management

✓ describe the different views held by individuals, organisations and governments on the management and sustainable use of water resources

✓ understand how the UK has attempted to manage its water resources in a sustainable way

✓ understand how China has attempted to manage its water resources in a sustainable way.

ACTIVITIES

1. Which households use the most water, metered or unmetered?
2. Give a reason for your answer to question 1.
3. Which large city in the Midlands does the Elan Valley reservoir supply with water?
4. In groups of four, imagine that you are either for or against the sustainable management of water. Devise a short play that highlights the arguments for or against the sustainable use of water.
5. Visit the following website. Create a poster giving advice on how to save water: www.how-to-save-water.co.uk

Extension

There is a debate about the use of meters to measure water usage. Do you think that water meters are a good idea? Give reasons for your answer.

Examination-style questions

1. Study Figure 13.25 on page 219.
   a) State the percentage of total usage of water used by public water supply in 2002. (1 mark)
   b) State in which year the most water was used for electricity supply. (1 mark)
   c) Calculate the difference in percentage water usage between electricity supply and public water supply in 2009. (1 mark)
2. Explain **one** reason for the increase in water demand per person in the past 50 years. (2 marks)
3. Suggest reasons for the changes to domestic water usage shown in Figure 13.25. (3 marks)
4. Explain the views of **two** different stakeholders, one in favour of the exploitation of water resources and the other one against it. (4 marks)
5. Assess how successful China has been in solving its water supply problems. (12 marks)

Total: 24 marks

Part 3 Geographical Investigations

In the following chapters, you will study the content you need for Component 3: Geographical Investigations.

Topic 7 Geographical Investigations: Fieldwork

In this topic you will study **Chapter 14:** an overview of investigating physical and human landscapes.

Topic 8 Geographical Investigations: UK Challenge

In this topic you will study **Chapter 15:** an overview of the physical and human characteristics of the UK.

What questions can I ask?

Should I break down my hypothesis / question into smaller questions which can be answered more easily?

What techniques am I going to use to collect the data?

Have I used observation to collect fieldwork evidence?

Have I used a range of techniques and methods to collect data in the field?

How will I explain and analyse my data?

Have I explained the data I collected in the field?

Have I added information from relevant case studies and theories to help me to explain my data?

How am I going to process and present my fieldwork data?

Have I used GIS, maps, graphs and diagrams (hand drawn and computer-generated to present and process my data?

How will I conclude and summarise my information?

Have I written logical conclusions based on evidence from my fieldwork notebook and the data I collected?

How well did I do?

Have you evaluated your fieldwork data and methods used?

Have conclusions been drawn?

What knowledge have you gained?

THE ROUTE TO ENQUIRY

Data

14 Geographical Investigations: Fieldwork

As part of your course you must complete two geographical enquiries. Both of these require you to carry out work in the field and research in the classroom. You must complete one piece of physical geography fieldwork, either in a river or coastal environment, and one piece of human geography fieldwork, either in a central/inner-city area or a rural settlement. Both pieces of work should be written up as an enquiry. You will then have to use your experiences and results to answer questions on an examination paper from memory.

KEY TERMS

Quantitative techniques – data collection techniques that record statistical data and/or measurements and are carried out in the field.

Qualitative techniques – techniques in which information is gained through observation; it usually involves a description of a feature.

Primary data source – first-hand evidence collected by the researcher themselves.

Secondary data source – evidence collected by someone else.

Catchment area – area from which a river drains water.

Discharge – the amount of water passing a specific point at a given time, measured in cubic metres per second. Calculate by cross-section area x velocity.

■ LEARNING OBJECTIVE

To study the requirements of geographical investigations in physical geography environments – river landscapes.

Learning outcomes

▶ To develop an understanding of the kinds of questions that can be investigated through fieldwork in river environments with particular emphasis on location and the task.

▶ To study different data collection techniques, including a quantitative fieldwork method to measure river discharge and a qualitative fieldwork method to record landforms that make up the river landscape.

▶ To be able to explain the implications of river processes for people living in the catchment area.

▶ To have experience of secondary sources of data such as the British Geological Survey's Geology of Britain viewer and one other secondary source of information.

River landscapes fieldwork

How to decide on your question(s)?

The place that you are going to complete your study will probably be decided on by your teacher because of the health and safety implications. The task that you will be completing is 'an investigation of change in a river channel', which has been set by the examination board.

Your job is to come up with a question or questions to answer which will relate to the task. In order to decide on the questions that you are going to be answering in the field, you will need to study the area you will be visiting very carefully and refer to your work on river landscapes to ensure that you know the processes that are involved. It is a good idea to have a question for each of the processes/river features that you will be investigating; examples are given in Figure 14.1.

Process/feature	Question
Gradient	Does the gradient of the river channel increase with distance from the source?
Depth and width	Do the depth and width of the river channel increase with distance from the source?
Velocity	Does the river velocity increase with distance from the source?
Bedload size and shape	Does the bedload size and shape change with distance from the source?
Wetted perimeter	Is the wetted perimeter greater at site 1 or site 5?
Cross-sectional area	Where is the cross-sectional area greatest?
Discharge	Does the discharge of the river increase with distance from the source?
Flooding	How have you defended yourself and your property against flooding?

⬆ **Figure 14.1** Possible questions for river environment studies.

What methods are you going to use to collect your data?

There are many different methods that can be used to collect information about river landscapes. They can be split into **primary data** collection, which can be **quantitative** or **qualitative**, and **secondary data** collection, which could be from the British Geology Survey's Geology of Britain viewer (www.bgs.ac.uk/discoveringgeology/geologyofbritain/viewer.html) or another source. You will also need to recognise different types of fieldwork equipment, such as clinometers and flow meters.

Methods for collecting quantitative data through fieldwork

Quantitative fieldwork techniques measure variables in the river environment. These methods are described in Figure 14.2.

Gradient – a clinometer is used to measure the gradient. A 5 m section of the river is measured out and marked with two ranging poles. The measurement is taken between the two poles and the degrees are noted. If this is completed at each site, the gradient of the long profile of the river can be calculated.

Wetted perimeter – this measurement is being carried out in both of these photographs. A chain is laid along the bed of the river and up the banks to the height of the water. The chain is then measured and the distance recorded. This technique can also be recorded using a tape measure instead of a chain, with people standing on the tape measure. It can also be calculated mathematically from width and depth measurements.

A number of techniques can be seen on this photograph.

Channel width – keep a tape measure just above the current level of the water. Measure from one bank to the other.

River depth – put a tape measure across the river. Take measurements at regular intervals across the river channel using a metre stick or a ranging pole and tape measure.

These two measurements can then be used to calculate the cross-section area of the river.

Bedload size and shape – bedload size is measured using a calliper. The rock is placed in the calliper and a measurement is read off the side. The shape is determined using Powers roundness index.

Velocity – measure 10 m along the river and place a ranging pole at each end. Drop an orange into the water before the first pole. Start timing as the orange passes the first pole and stop timing when it passes the second pole. Complete this ten times at different places across the river and take the average measurement.

⬆ **Figure 14.2** Methods for collecting quantitative data through fieldwork.

Methods for collecting qualitative data through fieldwork

Qualitative methods are where information is gathered using observation, so it reflects people's opinions of an area. It usually involves a description or comment about somewhere. These methods can be used to record landforms that make up the river landscape of an area. This type of data is collected using techniques such as photographs, sketches or questionnaires.

Photographs and sketches

Photographs are taken in the field of the area under discussion. Descriptions are put on to the photographs back in the classroom. Photographs show a more accurate view of an area but students must ensure that they record the annotations they want to add later. It is often a good idea to sketch and photograph the same area.

Sketches are drawn in the field with annotations and are usually then redrawn in the classroom to improve the presentation. These techniques can be used to provide visual evidence of how river processes impact on people living in an area.

Site 1: Afon Tarell river study

The river is near its source in the mountains.

These students are measuring the width of the river.

The channel has quite steep sides.

These students are using ranging poles to measure velocity.

The channel is small and the river is shallow and narrow because this is the upper course of the river.

⬆ **Figure 14.3** Annotated photograph of river study area.

Questionnaires

Another qualitative technique is a questionnaire. This can have closed or open questions, but all of them require the respondent to make a judgement or give an opinion on something. The evidence collected is therefore subjective. A questionnaire can be compiled and completed individually or in groups. It is a good way to find out information on how river processes such as flooding impact on the people living in an area.

Secondary data sources

The British Geology Survey's Geology of Britain viewer could be used to investigate the rock types of the study area. This information might be needed to investigate why the changes have occurred in the river channel. Other secondary sources, such as newspaper articles, could be used to find out information on floods in the area to assess the impact of river processes on the local people. Flood risk maps provided by the Environment Agency could also be used to assess the impact of the river on people who live in the area.

How are you going to present your data?

Your work will not be assessed as part of your examination but it is important that you have experienced the different ways that data can be presented. This is because you could be asked to display information in the examination or be asked to comment on the appropriateness of different presentation techniques.

Figure 14.4 lists the most common techniques that can be used to present data collected in river environments.

Technique	Example
Cartographic	A map of the study area
Graphical, simple	Bar charts
	Line graphs
	Pie charts
Graphical, sophisticated	Choropleth maps
	Flow line maps
	Scattergraphs
Visual	Photographs
	Field sketches
Analytical calculation	Work out the cross-sectional area and discharge (see Figure 14.5)

⬆ **Figure 14.4** Data presentation techniques.

How to calculate cross-sectional area and river discharge

Cross-sectional area

This is calculated by using width and depth measurements. There are two ways to do this: either multiply the width by the depth (you may need to find the mean of the depth measurements first), or plot the width and depth measurements on to graph paper, as shown in Figure 14.5, keeping the same scale for each of the axes. The squares where the water is shown are then counted to calculate the measurement, which is in m^2.

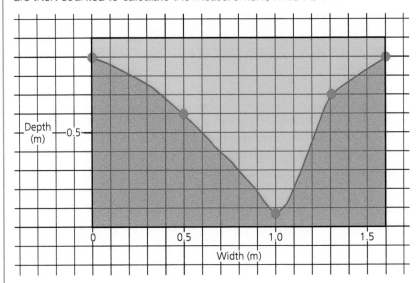

⬆ **Figure 14.5** Cross-sectional area.

River discharge

Discharge is calculated by multiplying the cross-sectional area by the mean velocity for each site. For example, if the mean velocity is 0.95 metres per second and the cross-sectional area is 1.50 m^2:

Discharge = 0.95 × 1.50 = 1.42 m^3 per second (or cumecs)

Review

By the end of this section you should be able to:

✓ develop questions that can be answered in fieldwork environments
✓ understand quantitative and qualitative fieldwork techniques, in particular those used to measure river discharge and to record landforms that make up the river landscape
✓ explain the implications of river processes for people living in the catchment area
✓ recognise secondary data sources, such as the British Geology Survey's Geology of Britain viewer.

Coastal landscapes fieldwork

To study the requirements of geographical investigations in physical geography environments – coastal landscapes.

Learning outcomes

▶ To develop an understanding of the kinds of questions that can be investigated through fieldwork in coastal environments with particular emphasis on location and the task.
▶ To study different data collection techniques including a quantitative fieldwork method to measure beach morphology and sediment characteristics, and a qualitative fieldwork method to record landforms that make up the coastal landscape.
▶ To be able to explain the implications of coastal processes for people living in the catchment area.
▶ To have experience of secondary data sources such as the British Geological Survey's Geology of Britain viewer and one other secondary source of information.

How to decide on your question(s)?

The place that you are going to complete your study will probably be decided on by your teacher because of the health and safety implications. The task that you will be completing is '*an investigation of coastal processes through landscape evidence*', which has been set by the examination board.

Your job is to come up with a question or questions to answer which will relate to the task. In order to decide on the questions that you are going to be answering in the field, you will need to study the area you will be visiting very carefully and refer to your work on coastal landscapes to ensure that you know the processes that are involved. It is a good idea to have a question for each of the processes/coastal features that you will be investigating; examples are given in Figure 14.6.

Coastal feature	Question
Sediment characteristics – size and shape of pebbles	Does the sediment size and shape change with distance from the cliff face?
Beach profiles	Is the beach steeper closer to the cliff face?
Longshore drift	Is longshore drift occurring on this coastline?
Groyne sediment	Do the groynes have a build-up of sediment on one side?
Cliff profile	Are the cliffs steeper in the west of the beach than the east?

⬆ **Figure 14.6** Possible questions for coastal environment studies.

What methods are you going to use to collect your data?

There are many different methods that can be used to collect information about coastal landscapes. They can be split into primary data sources, which can be quantitative or qualitative methods, and secondary data sources, which could be from the British Geology Survey's Geology of Britain viewer (www.bgs.ac.uk/discoveringgeology/geologyofbritain/viewer.html) or past data/information on the area. You will also need to recognise different types of fieldwork equipment, such as quadrats and callipers.

Methods for collecting quantitative data through fieldwork

Quantitative fieldwork techniques measure variables on beaches.
These methods are described in Figure 14.7.

Beach morphology measured through beach profiles – these are completed using ranging poles and a clinometer. Start by the sea and measure the angle between two ranging poles every 10 m until the cliff face is reached.

Measuring the speed and direction of longshore drift – an orange is thrown into the sea. A ranging pole is placed on the beach close to the sea at this point and a stopwatch is started. It was stopped after five minutes had passed. The distance the orange had travelled was then measured with a tape measure. The test was repeated five times to get an average speed.

Sediment characteristics, the size and shape of pebbles – this is usually completed with a beach profile. A quadrat is placed on the ground (if it is not available two metre rules will be fine) where each ranging pole is placed. Ten stones are picked, two from each corner and two from the middle. They are measured with a calliper and their shape is checked against Powers roundness index.

Measuring the build up of beach material beside groynes to check for longshore drift – the depth of sand is measured on either side of the groyne. This is done at two points along the groyne, in the middle of the groyne and next to the sea wall. This is repeated on all the groynes on the beach.

⬆ **Figure 14.7** Quantitative data collection methods.

Methods for collecting qualitative data through fieldwork

Qualitative methods are where information is gathered using observation, so it reflects people's opinions of an area. It usually involves a description or comment about somewhere. These methods can be used to record landforms that make up the coastal landscape of an area. This type of data is collected using techniques such as photographs, sketches or questionnaires.

Photographs and sketches

Photographs are taken in the field of the area under discussion. Descriptions are put on to the photographs back in the classroom. Photograph show a more accurate view of an area but students must ensure that they record the annotations they want to add later. It is often a good idea to sketch and photograph the same area.

Sketches are drawn in the field with annotations but are usually redrawn in the classroom to improve the presentation. These techniques could be used to provide visual evidence of how coastal processes impact on people living in an area.

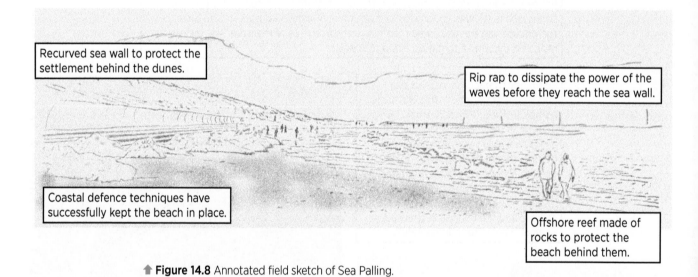

Recurved sea wall to protect the settlement behind the dunes.

Rip rap to dissipate the power of the waves before they reach the sea wall.

Coastal defence techniques have successfully kept the beach in place.

Offshore reef made of rocks to protect the beach behind them.

⬆ **Figure 14.8** Annotated field sketch of Sea Palling.

Questionnaires

Another qualitative technique is a questionnaire. This can have closed or open questions, but all of them require the respondent to make a judgement or give an opinion on something. The evidence collected is therefore subjective. A questionnaire can be compiled and completed individually or in groups. It is a good way to find out information on how coastal processes such as erosion impact on the people living in an area.

Secondary data sources

The British Geological Survey's Geology of Britain viewer could be used to investigate the rock types of the study area. This information might be needed to investigate why changes have occurred to the coastal landscape. Other secondary sources, such as newspaper articles, could be used to find out information on coastal recession in the area to assess the impact of coastal processes on the local people. Students could also use old maps of the area which could be found online or at the local library.

How are you going to present your data?

Your work will not be assessed as part of your examination but it is important that you have experienced the different ways that data can be presented. This is because you could be asked to display information in the examination or be asked to comment on the appropriateness of different presentation techniques.

Figure 14.9 lists the most common types techniques that can be used to present data collected in coastal environments.

Technique	Example
Cartographic	A map of the study area
Graphical, simple	Bar charts Line graphs Pie charts
Graphical, sophisticated	Choropleth maps Flow line maps Scattergraphs
Visual	Photographs Field sketches
Analytical	Spearman's rank correlation co-efficient for pebble size and shape survey

⬆ **Figure 14.9** Data presentation techniques.

Review

By the end of this section you should be able to:

✓ develop questions that can be answered in fieldwork environments
✓ understand quantitative and qualitative fieldwork techniques, in particular those to measure beach morphology and methods to record landforms that make up the coastal landscape
✓ explain the implications of coastal processes for people living in the catchment area
✓ recognise secondary data sources such as the British Geology Survey's Geology of Britain viewer.

Human landscapes fieldwork (central/inner urban area)

To study the requirements of geographical investigations in human geography environments – central/inner urban landscapes.

Learning outcomes

▶ To develop an understanding of the kinds of questions that can be investigated through fieldwork in urban environments with particular emphasis on location and the task.

▶ To study different data collection techniques including a quantitative fieldwork method to measure land use function and a qualitative fieldwork method to record the quality of the urban environment.

▶ To be able to understand the interaction between physical landscape features and the area's residents and visitors.

▶ To have experience of secondary data sources, such as the neighbourhood statistics from the Office for National Statistics and one other secondary source of information.

How to decide on your question(s)?

The place that you are going to complete your study will probably be decided on by your teacher because of the health and safety implications. The task that you will be completing is 'investigating change in central/inner urban area(s)', which has been set by the examination board.

Your job is to come up with a question or questions to answer which will relate to the task. In order to decide on the questions that you are going to be answering in the field you will need to study the area you will be visiting very carefully and refer to your work on central and inner urban areas to ensure that you know the changes that have taken place. Figure 14.10 gives examples of questions that you could ask about the central or inner urban area.

Urban feature	Question
Land-use function	How does land use change with distance from the city centre?
Land-use change	How has the land use in the CBD changed since the Second World War?
Quality of environment	How does quality of the environment change with distance from the city centre?
Physical landscape	How has the river impacted on the land use and the people who use the central/inner urban area?

⬆ **Figure 14.10** Possible questions for urban environment studies.

What methods are you going to use to collect your data?

Many different methods can be used to collect information about urban areas. They can be split into primary data sources, which can be quantitative or qualitative methods, and secondary data sources, which could be statistics from the Office for National Statistics or another source, such as GOAD maps.

KEY TERMS

Stratified sampling – the population is divided into sections known as 'strata'; the sampler ensures that the same amount of data is taken from each stratum. For example, the strata could be roads in the CBD. The sampler would ensure that they interviewed ten people on each road.

Random sampling – this is where things are chosen by chance. A random number table could be used.

Systematic sampling – this means the sample is chosen according to an agreed interval. For example, every fifth person who walks passed is interviewed.

GOAD maps – detailed street maps including individual buildings and their uses.

Methods for collecting quantitative data through fieldwork

Quantitative fieldwork techniques are techniques that measure variables such as the different land use in the central city or the numbers of different types of shops. Students could also measure shop frontage to see if this changes with distance from the centre of the city.

Land-use survey

A transect is carried out down streets leading from the centre of the city. Students walk down a street noting down the land use function of the buildings on a map using the key shown in Figure 14.11.

Information is shared and numerous data handling and analysis techniques can be carried out on the information collected.

Land use	Code
Residential	R
Industrial	I
Commercial (including retail)	C
Entertainment	E
Public buildings	P
Open space	O
Transport	T
Services	S

⬆ **Figure 14.11** Land-use survey form.

Methods for collecting qualitative data through fieldwork

Qualitative methods are where information is gathered using observation, so it reflects people's opinions of an area. It usually involves a description or comment about somewhere. This type of data is collected using techniques such as photographs, sketches, questionnaires (which have been discussed earlier in this chapter) and environmental quality surveys.

Environmental quality survey

This involves the students giving a score to the quality of the environment in different areas of the CBD. Figure 14.12 shows an example of a survey sheet that could be used for this technique.

Negative	1	2	3	4	5	Positive
There is lots of litter						There is no litter
Noisy						Very quiet
Lots of dog fouling						No dog fouling
No hanging baskets or greenery						Lots of greenery
Shop fronts are run down and dirty						Shop fronts are newly painted and well maintained
Shop windows are dirty						Shop windows are clean
Lots of traffic						Pedestrianised area
Few shoppers						Many shoppers
Buildings are old and not cared for						Buildings are new and modern

⬆ **Figure 14.12** Environmental quality survey form.

The information collected could be shared between the students.

Secondary data sources

The Office for National Statistics can be used to get information on the people who live in the study area. This information might be needed to investigate why changes have occurred to the urban landscape. Other secondary sources, such as newspaper articles, could be used to find out information about what the urban area was like in the past. GOAD maps of the area could also be used to find out about previous land uses.

How are you going to present your data?

Your work will not be assessed as part of your examination but it is important that you have experienced the different ways that data can be presented. This is because you could be asked to display information in the examination or be asked to comment on the appropriateness of different presentation techniques.

Figure 14.13 lists the most common types of techniques used to present data collected in urban environments.

Technique	Example
Cartographic	A map of the study area
Graphical, simple	Bar charts
	Line graphs
	Pie charts
Graphical, sophisticated	Choropleth maps
	Flow line maps
	Scattergraphs
	Transects
	Proportional symbols
Visual	Photographs
	Field sketches

⬆ **Figure 14.13** Data presentation techniques.

Review

By the end of this section you should be able to:

✓ develop questions that can be answered in fieldwork environments
✓ reproduce quantitative and qualitative fieldwork techniques, in particular those to measure land-use function and to record the quality of the urban environment
✓ understand the interaction between physical landscape features and the residents and visitors to the central/inner urban area
✓ understand secondary data sources such as the neighbourhood statistics of the Office for National Statistics and one other secondary source of information.

Human landscapes fieldwork (rural settlements)

LEARNING OBJECTIVE

To study the requirements of geographical investigations in human geography environments – rural landscapes.

Learning outcomes

▶ To develop an understanding of the kinds of questions that can be investigated through fieldwork in rural environments with particular emphasis on location and the task.

▶ To study different data collection techniques including a quantitative fieldwork method to measure flows of people within a rural settlement and a qualitative fieldwork method to record the views of people on the quality of the rural environment.

▶ To be able to understand the interaction between physical landscape features, rural settlements, residents and visitors.

▶ To have experience of secondary data sources such as the neighbourhood statistics of the Office for National Statistics and one other secondary source of information.

KEY TERMS

Stratified sampling – the population is divided into sections known as 'strata'; the sampler ensures that the same amount of data is taken from each stratum. For example, the strata could be roads in the CBD The sampler would ensure that they interviewed ten people on each road.

Random sampling – this is where things are chosen by chance A random number table could be used.

Systematic sampling – this means the sample is chosen according to an agreed interval For example, every fifth person who walks passed is interviewed.

GOAD maps – detailed street maps including individual buildings and their uses.

How to decide on your question(s)?

The place that you are going to complete your study will probably be decided on by your teacher because of the health and safety implications. The task that you will be completing is '*investigating change in rural settlements*', which this has been set by the examination board.

Your job is to come up with a question or questions to answer which will relate to the task. In order to decide on the questions that you are going to be answering in the field, you will need to study the area you will be visiting very carefully. Figure 14.14 gives examples of questions that you could ask about rural areas.

Rural feature	Question
Closure of rural services	How has the closure of services in the village changed the village community?
Changes to the area caused by tourism	How has tourism impacted on the life of the village?
Physical landscape	How important is the physical landscape on the shape and character of the rural settlement?

⬆ **Figure 14.14** Possible questions for rural environment studies.

What methods are you going to use to collect your data?

Many different methods can be used to collect information about rural areas. They can be split into primary data sources, which can be quantitative or qualitative methods, and secondary data sources, which could be statistics from the Office for National Statistics or another source, such as fieldwork data from previous years.

Methods for collecting quantitative data through fieldwork

Quantitative fieldwork techniques are techniques that measure variables such as the flow of people or traffic through a rural settlement.

Pedestrian and traffic counts

Students count pedestrians/traffic at certain points in the rural settlement for fifteen minutes every hour. The information is then collated and flow maps are produced from the data.

Methods for collecting qualitative data through fieldwork

Qualitative methods are where information is gathered using observation, so it reflects people's opinions of an area. It usually involves a description or comment about somewhere. This type of data is collected using techniques such as photographs and sketches. If people's views on the quality of the rural environment are to be recorded, the best method is a questionnaire or an environmental quality survey. All of these techniques have been discussed earlier in this chapter.

Secondary data sources

The Office for National Statistics can be used to get information on the people who live in the study area. This information might be needed to investigate why changes have occurred to the urban landscape. Other secondary sources, such as newspaper articles, could be used to find out information about what the rural area was like in the past.

How are you going to present your data?

Your work will not be assessed as part of your examination, but it is important that you have experienced the different ways that data can be presented. This is because you could be asked to display information in the examination or be asked to comment on the appropriateness of different presentation techniques.

Figure 14.15 lists the most common techniques used to present data collected in urban environments.

Technique	Example
Cartographic	A map of the study area
Graphical, simple	Bar charts
	Line graphs
	Pie charts
Graphical, sophisticated.	Choropleth maps
	Flow line maps
	Scattergraphs
	Transects
	Proportional symbols
Visual	Photographs
	Field sketches

⬆ **Figure 14.15** Data presentation techniques.

Review

By the end of this section you should be able to:

✓ develop questions that can be answered in fieldwork environments
✓ reproduce quantitative and qualitative fieldwork techniques, in particular those to measure land-use, the flow of people and traffic, and to record the quality of the rural environment
✓ understand the interaction between physical landscape features, the rural settlement and residents and visitors
✓ understand secondary data sources such as the neighbourhood statistics of the Office for National Statistics and one other secondary source of information.

Data analysis, conclusions and evaluation

These must be completed for both of the enquiries that you carry out.

Data analysis

In the examination you may be asked to analyse data that has been given to you; therefore, you must ensure that you understand how to complete an analysis. One way of doing this is to analyse the data you have obtained through your own fieldwork. You may also be asked to write an analysis of your findings during your field study, so it is essential that an analysis has been completed.

Conclusions

You must write a concluding comment on your fieldwork. You could be asked in the examination about your conclusions or be asked to write a conclusion from information you have been given. When you write a conclusion, you should ensure that:

- all questions have been answered
- evidence from the data you have collected has been included
- if your study was based on a particular theory, that this theory has been referred to and proved or disproved.

Evaluation

You should always evaluate the work you have completed: this means assessing the value of it. In order to do this you should ask yourself the following questions.

1 Were the methods I used appropriate?
2 Did the methods I used help me to answer my question?
3 Was I able to answer my original question given the primary and secondary data I collected?
4 How could I improve my study?

Measures of central tendency

This section gives information on the different ways to work out an average.

Mean

Add all of the numbers together and divide the total by the amount of numbers.

Mode

This is the value that occurs most often in a set of observations.

Median

Put the numbers into numerical order. If there is an odd number of results, the middle result is the median. If there is an even number of results, the median is the mean of the two numbers in the centre of the numerical order.

Practise your skills

Work out the mean, mode and median for the river data in the table.

Site	River width (m)
1	1.5
2	3
3	2.5
4	4
5	4
6	4.8
7	5.6
8	6.8
9	6.5
10	7
11	7.5

Examination-style questions

Investigating physical environments (rivers)

1 You have carried out fieldwork in a river environment.
  Name your river environment _____
  a) Explain **one** primary data collection technique you used in the field.
     Name of primary data collection technique _____ (3 marks)
  b) Define the term quantitative data collection techniques. (1 mark)
  c) During your fieldwork you will have studied secondary data sources.
     Name of secondary data collection method _____
     Explain how the information you discovered helped to answer your fieldwork question. (3 marks)
  d) Explain **one** way that river processes in the area of your study have had an impact
     on the local people. (3 marks)
  e) A group of students were investigating a river in Wales. Figure 14.16 is an extract
     from one student's fieldwork write-up.

We were studying the changes in a river channel as it moves from its source to its mouth. We chose five sites randomly along the river, although one was near the source and the last was near the mouth.

We measured the velocity by placing a ranging pole in the river measuring five metres and then placing another one in. One person then threw an orange in the river by the ranging pole and we timed how long it took for it to float past the other pole. We measured the depth with a tape measure at the banks and in the middle. We measured the width by holding the tape measure tightly between the two banks. We measured the gradient with a clinometer. The results we obtained showed that the river got wider and deeper as it moved downstream, and it also started to move more quickly. The gradient became less steep as we moved downstream.

⬆ **Figure 14.16** Extract from a student's river environment fieldwork write-up.

        Evaluate the student's methodology. (8 marks)
                                                                          Total: 18 marks

Investigating physical environments (coasts)

2 You have carried out fieldwork in a coastal environment.
  Name your coastal environment _____
  a) Explain **one** primary data collection technique you used in the field.
     Name of primary data collection technique _____ (3 marks)
  b) Define the term quantitative data collection techniques. (1 mark)
  c) During your fieldwork you will have studied secondary data sources.
     Name of secondary data collection method _____
     Explain how the information you discovered helped to answer your fieldwork question. (3 marks)
  d) Explain **one** way that coastal processes in the area of your study have had an impact
     on the local people. (3 marks)
  e) A group of students were investigating a coastal area in Dorset. Figure 14.17 is an extract
     from one student's fieldwork write up.

We were studying the processes that occur on a beach. We chose five sites randomly along the beach.

We placed a ranging pole on the beach and measured ten metres to place another one. Then we timed how long it took for an orange to move between the poles. We then did a beach profile using ranging poles, a tape measure and a clinometer. This allowed use to see the change in gradient as we moved from the sea to the cliff. We did this when there was a break in slope.

At the same place as we put are ranging poles in, we also used our quadrat to help use choose ten pebbles, which were chosen at random. We measured them and compared them to Powers roundness index to check their shape.

We found that the oranges always moved from south to north, the beach got steeper as we moved towards the cliff, and that the pebbles were bigger nearer to the cliff and also more angular.

⬆ **Figure 14.17** Extract from a student's coastal environment fieldwork write-up.

Evaluate the student's methodology. (8 marks)

Total: 18 marks

Changes in the central/urban area/CBD

3 Figure 14.18 shows a survey sheet used by a group of students to investigate the environment quality of an urban area.
  a) State **one** disadvantage of this technique. (1 mark)

Category	1	2	3	4	5	Category
Ugly						Beautiful
Noisy						Quiet
All buildings						Green
Dirty and littered						Clean and tidy
Closed in						Spacious
Danger from traffic						Traffic-free
Far from amenities						Close to amenities
Lots of pollution						Little pollution
Old fashioned						Modern
Haphazard						Variety of buildings
I don't like it						I like it
Hostile						Friendly

⬆ **Figure 14.18** Environment quality survey urban environments.

b) Figure 14.19 shows the average results for each site. The maximum score was 55.

Location	Person 1	Person 2	Person 3	Person 4	Person 5	Mean
1	42	39	47	44	39	42.2
2	36	40	37	35	42	38
3	42	44	42	43	38	
4	45	40	43	48	38	42.8
5	38	35	34	40	30	35.4

⬆ **Figure 14.19** Environment quality survey urban environments results.

    i) Calculate the mean score for location 3.   (1 mark)

    ii) Calculate the modal score for location 1.   (1 mark)

    iii) Suggest reasons why a student might use the mean score rather than the modal score when analysing their results.   (3 marks)

c) The students measured the distance between each location by pacing out 50 m. This is known as systematic sampling.
Suggest reasons why they used this sampling method.   (3 marks)

d) Pacing out a distance is not a reliable way of completing fieldwork. Suggest a piece of equipment that could have been used to make their results more reliable.   (1 mark)

e) You have carried out this type of fieldwork in an urban area.
Evaluate the methods that you used and the results that you obtained.   (8 marks)

Total: 18 marks

Changes in rural settlements

4   Figure 14.20 shows a survey sheet used by a group of students to investigate the environmental quality of a rural area.

Category	1	2	3	4	5	Category
Ugly						Beautiful
Noisy						Quiet
All buildings						Green
Dirty and littered						Clean and tidy
Closed in						Spacious
Danger from traffic						Traffic-free
Far from amenities						Close to amenities
Lots of pollution						Little pollution
Old fashioned						Modern
Haphazard						Variety of buildings
I don't like it						I like it
Hostile						Friendly

⬆ **Figure 14.20** Environment quality survey rural environments.

a) State **one** disadvantage of this technique.   (1 mark)

5 Figure 14.21 shows the average results that they obtained. The maximum score was 55.

Location	Person 1	Person 2	Person 3	Person 4	Person 5	Mean
1	44	42	50	44	48	45.6
2	36	42	39	37	44	39.6
3	42	44	42	43	42	
4	50	42	46	42	46	45.2
5	51	45	48	46	48	47.6

⬆ **Figure 14.21** Environment quality survey rural environment results.

  i)   Calculate the mean score for location 3.   (1 mark)
  ii) Calculate the modal score for location 1.   (1 mark)
  iii) Suggest reasons why a student might use the mean score rather than the modal score
       when analysing their results.   (3 marks)
  c) The students measured the distance between each location by pacing out 10 m.
      This is known as systematic sampling.
      Suggest reasons why they used this sampling method.   (3 marks)
  d) Pacing out a distance is not a reliable way of completing the fieldwork. Suggest a piece of
      equipment that could have been used to make their results more reliable.   (1 mark)
  e) You have carried out this type of fieldwork in a rural area.
      Evaluate the methods that you used and the results that you obtained.   (8 marks)
                                                                               Total: 18 marks

15 Geographical Investigations: UK Challenge

In this topic you will be required to draw on the knowledge and understanding you have gained from Chapters 1 to 13. You will also have to draw on your geographical skills to investigate a contemporary challenge facing the UK. The UK challenge will be on one or more of the four themes that will be dealt with in this chapter.

Theme 1: The UK's use of resources and environmental sustainability

UK challenge	Information you have already learnt	Where to find information in this book	
		Chapter	Page
Increasing population and use of resources	How human activities produce greenhouse gases	5	67
	UK variety and distribution of natural resources	6	96
Increasing population and pressures on the UK's ecosystems	Distribution and characteristics of the UK's main terrestrial ecosystems	6	95
	Importance of marine ecosystems to the UK as a resource and how human activities are degrading them	6	97
	Causes of national and international migration	8	130
	The ways in which people exploit environments to obtain water, food and energy	11	176
	How environments are changed by this exploitation	11	177
Sustainable transport options for the UK	Developments in transport causing deindustrialisation	8	134
	The range of sustainable transport options for your chosen UK city	8	137

⬆ **Figure 15.1** UK population and resources challenge.

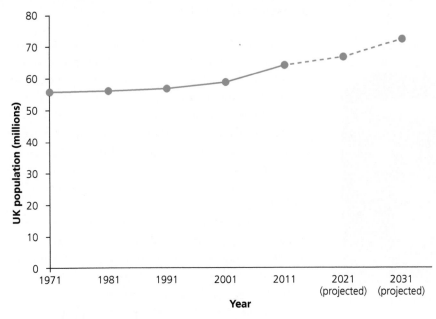

⬆ **Figure 15.2** The population of the UK, 1971–2031.

Changes to the UK's population during the next 50 years and implications for the use of resources

The UK's population has grown considerably since the year 2000 and projections show that it will continue to grow over the next 50 years. This growth will have an implication on resources. People will need more food to eat, water to drink and they will use more energy, which will put pressure on resources. The page references to where this is dealt with in more detail are given in Figure 15.1.

Pressures on the UK's ecosystems of the growing population

This challenge relates to the fact that the population of the UK is predicted to grow during the next 50 years and focuses on the impact of this on the ecosystems in the UK. The more people that live in the country, the greater the pressure there is not only to supply the population with the food that it requires but also to provide homes and transport routes. This involves the removal of vegetation, and therefore ecosystems, to provide homes for people. The page references to where this is dealt with in more detail are given in Figure 15.1 on page 242.

Different sustainable transport options used in the UK

If the population of the UK grows at its present rate, cities and towns will have to provide more sustainable transport options because of the problems of not enough space for all the private cars on the road. Chapter 8 looks at this in some detail for Bristol. Figure 15.3 shows all the sustainable transport options that are available for UK cities.

↑ **Figure 15.3a** Sustainable transport options in the UK: park and ride for Durham.

↑ **Figure 15.3b** Sustainable transport options in the UK: congestion charging zone, Durham.

← **Figure 15.3c** Sustainable transport options in the UK: trams in Sheffield, an alternative to buses.

Sustainable transport options:

- Car sharing where workers share lifts to work using their own cars. If half of the UK motorists received a lift one day a week, vehicle congestion and pollution would be reduced by ten per cent and traffic jams by twenty per cent.
- Designated cycle and walking paths within the urban area. Milton Keynes is one of the best served urban areas in the UK with 273 km of cycle paths.
- Road lanes that only allow cars with at least two passengers to use them.
- Pedestrianised areas which restrict private vehicle access but allow buses and trams to operate.
- Road lanes that give priority to buses, ensuring they get an easy passage through congested areas.
- Restricting car parking in central urban areas so motorists are forced to use public transport.
- Park and ride schemes, which allow shoppers to park their cars in large designated parking areas on the edge of the urban area and catch a bus into the town centre. Park and ride operates in 87 towns and cities in the UK. Parking is free but there is often a charge for bus travel to the city centre. The park and ride sites are usually located on the main routes coming into the urban area, so they are easily accessible for the greatest number of car users.
- Congestion charging, which is the practice of making motorists pay to travel into large urban areas during periods of heaviest use. The aim is to reduce the number of vehicles entering the city, which eases traffic congestion and lowers pollution emissions. It will hopefully lead to more sustainable forms of transport such as walking, cycling or public transport being used.

Theme 2: The challenges presented by UK settlement, population and economic changes

To study the challenges presented by UK settlement, population and economic changes.

Learning outcomes

▶ To understand the 'two-speed economy' and the options for bridging the gap between the South East and the rest of the UK.
▶ To know the costs and benefits of greenfield development and the regeneration of brownfield sites.
▶ To know UK net migration statistics and their reliability and values, and attitudes of different stakeholders towards migration.

KEY TERMS

Two-speed economy – the South East has a faster rate of growth than the rest of the UK.

Stakeholder – someone with an interest in what occurs.

UK challenge	Information you have already learnt	Where to find information in this book	
		Chapter	Page
The two-speed economy and options for bridging the gap between the South East and the rest of the UK	Factors causing the rate and degree of urbanisation to differ between the regions of the UK	7	121
Costs and benefits of greenfield development and the regeneration of brownfield sites	Recent changes in retailing and its impact on UK cities	8	135
	The sequence of urbanisation, suburbanisation, counter-urbanisation and re-urbanisation processes and their distinctive characteristics	8	129
UK net migration statistics and their reliability and values, and attitudes of different stakeholders in migration	Causes of national and international migration and its impact on UK cities	8	130

⬆ **Figure 15.4** Challenges presented by UK settlement, population and economic changes.

The two-speed economy and options for bridging the gap between the South East and the rest of the UK

The difference between the growth of the South East of the UK and the rest of the country is becoming more marked. Many companies prefer to have their headquarters in this area, especially in London, because there are better transport routes there to both Europe and the rest of the world. They also believe that firms located in London are thought of in a better light by other companies. This has led to wages in London generally being higher than the rest of the country. House prices and the cost of living are also higher. This is a negative effect of the two-speed economy for a lot of people who live in the area.

Some companies have moved their head offices away from the capital because of the expensive housing and living costs. For example, in the 1990s Lloyds Bank moved its head office to Bristol and HSBC moved to Birmingham. The government has also moved some of its key departments out of the South East, such as the Passport Office which is in Cardiff.

The government is also trying to create link between the large cities in the North through rail and road links to encourage companies to locate there. It is also building a high-speed train link to the Midlands and North of the UK. The belief is that if people can travel swiftly to and from London then companies are more likely to locate out of the South East.

Costs and benefits of greenfield development and building on brownfield sites

Greenfield development sites are found on the edge of cities and have never been built on before. The costs and benefits of these developments can be seen in Figure 15.5. Brownfield sites are areas in cities that have been built on before and could contain old buildings. The local councils and the government are keen for this land to be developed to stop cities from spreading any further into the countryside. There are, however, costs and benefits of this development, as outlined in Figure 15.6.

Benefits	Costs
Originally unoccupied therefore developers can build as they wish.	Infrastructure such as gas, electricity and water will not be present.
Plenty of space for car parking and landscaping to improve the working environment.	Urban sprawl, using up green spaces on the edge of urban areas, disturbs natural habitats and wildlife.
Cheaper land due to being further from the city centre.	It is more difficult to get planning permission as the government tends to be against it.
Lower construction costs as there is nothing to knock down or renew.	Living on the edge of the city may increase the commute for some people.
Easy to market to potential buyers because of the pleasant environment.	Disruption to the local area during construction.
Access to the development is easier as roads are not congested.	People may not want to live away from the city centre because of their social life.

⬆ **Figure 15.5** Benefits and costs of developing greenfield sites.

Benefits	Costs
It is easy for the company to get planning permission because the government is actively encouraging the use of these sites.	Complete environmental survey needed because of past usage, which is costly and time consuming.
The infrastructure, such as power and water, is already present so there is no disruption to the system.	Perception of contaminated environment puts off prospective buyers.
Greenfield sites are not used, so lessens urban sprawl.	Cities may have social problems – such as antisocial behaviour and crime – as well as higher levels of pollution and congestion, which could make marketing more difficult.
Development costs are less as much of the infrastructure and roads are already present.	Land costs are higher as it is closer to the city centre.

⬆ **Figure 15.6** Benefits and costs of developing brownfield sites.

⬆ **Figure 15.7** A brownfield site in Leeds awaiting redevelopment.

UK net migration statistics and their reliability and values, and attitudes of different stakeholders towards migration

Net migration statistics show the difference between the number of people who are coming into a country and the number of people leaving a country. In the UK the number is rising as more people from other countries in Europe and the rest of the world wish to live in the UK. The reliability of these numbers can be a problem, both to people who produce the statistics and to the government. The government only has records of people who come into the country legally. It is believed that many people come into the country without being recorded and without permission. This makes the figures inaccurate and leads to other problems such as these people working for very little pay because they should not be in the country. Their employers know this and exploit the situation. There are many different viewpoints about the topic of migration, some of which are suggested in Figure 15.9 on page 248.

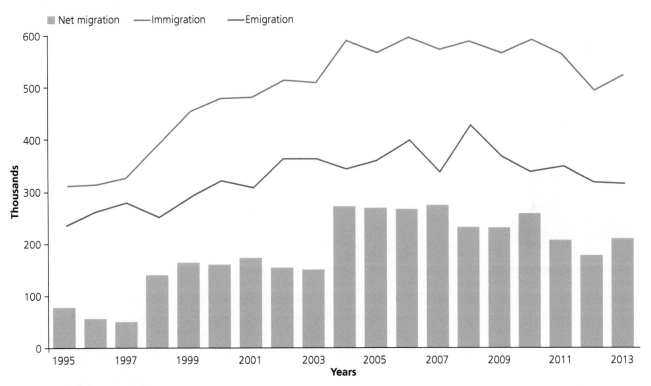

⬆ **Figure 15.8** Net migration figures for the UK.

There are more people living in the town so I sell more goods. Migrant workers earn an average of £20,000 per year of which £6,000–£7,000 is disposable income.

Local shopkeeper

They bring new cultures and traditions, which adds to the cultural mix of British society. I love being able to buy a great variety of food in my local shop.

Local resident

They are good for the country because they are prepared to do jobs that British workers are not, for example working on farms and in farming-related industries in Cambridgeshire.

Government spokesperson

The increase in population numbers has caused problems for the NHS as it cannot cope with the number of people needing its services. We have also seen a large increase in claims for child benefit.

Government spokesperson

In Cambridgeshire the police force has to deal with 100 different languages. This has cost £800,000 to pay for translators. We now employ community support officers to translate for us.

Local police officer

⬆ **Figure 15.9** The values and attitudes of different stakeholders towards migration.

Theme 3: The UK's landscape challenges

Learning outcomes

▶ To understand the approaches to conservation and development of the UK's national parks.
▶ To know the approaches to managing the UK's rivers and coasts.

KEY TERMS

National park – an area of countryside that is protected because of its natural beauty and managed for visitor recreation.

Conservation – keeping something as it is; not changing it in any way to preserve it for future generations.

UK challenge	Information you have already learnt	Where to find information in this book	
		Chapter	Page
To understand the approaches to conservation and development of the UK's national parks	How human activities can change upland areas	4	52
	The advantages and disadvantages of development in upland areas	4	53
To know the approaches to managing the UK's river and coastal	The different defences used on the coastline of the UK	2	18–22
	The different defences used on rivers in the UK	3	38

⬆ **Figure 15.10** The UK's landscape challenges.

Approaches to conservation and development of the UK's national parks

National parks are areas of land that have outstanding value in terms of their natural beauty, environment or recreational value. The designation as a national park gives the area special protection and means that resources are available to promote and manage tourism. Special funds are available to landowners and certain restrictions apply on development. Much of the land in the UK's national parks is owned by private individuals; it is not publically owned as it is in other countries. The National Trust owns some of the land and therefore has a say in the management of these areas. Each of the national parks are run by their own National Park Authority. The authorities have two statutory purposes:

1 To conserve and enhance the natural beauty, wildlife and cultural heritage of the area.
2 To promote opportunities for the understanding and enjoyment of the park's special qualities by the public.

The main power of the National Park Authority is to control development. Their funding is mainly from central government. The parks employ a number of people including rangers who are responsible for conserving and enhancing the natural beauty, wildlife and cultural heritage of the park, as well as improving opportunities for the public to use the park. If there is any conflict between conservation and enjoyment, conservation always takes priority. Each of the national parks has its own management plan which states the objectives of the park and the ways in which these objectives are to be met; it is the overarching vision of the park. The plans last for five to ten years and are derived through consultation with all the stakeholders, including local authorities, the National Trust and local landowners.

🔺 **Figure 15.11** The Yorkshire Dales and Lake District national parks.

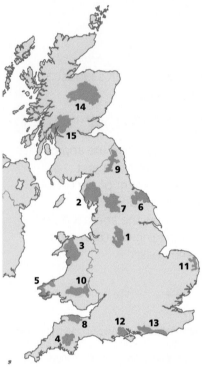

Key	National Park	Established	Area (km²)
1	Peak District	1951	1 438
2	Lake District	1951	2 292
3	Snowdonia	1951	2 142
4	Dartmoor	1951	956
5	Pembrokeshire	1952	620
6	North York Moors	1952	1 436
7	Yorkshire Dales	1954	1 769
8	Exmoor	1954	693
9	Northumberland	1956	1 049
10	Brecon Beacons	1957	1 351
11	The Broads	1988	303
12	New Forest	2005	580
13	South Downs	2008	1 641
14	Cairngorms	2003	3 800
15	Loch Lomond & The Trossachs	2002	1 865

🔺 **Figure 15.12** National parks in the UK.

Approaches to managing river and coastal UK

These topics are covered in depth in the specification. Different coastal and river defence techniques are discussed with their advantages and disadvantages; see Figure 15.10 on page 249 for the relevant page numbers.

Theme 4: The UK's climate change challenges

To study the UK's climate change challenges.

Learning outcomes

▶ To know that there are uncertainties about how global climate change will impact on the UK's future climate.
▶ To understand the impacts of climate change on people and landscapes in UK.
▶ To recognise that there are a range of responses to climate change in the UK at a local and national scale.

UK challenge	Information you have already learnt	Where to find information in this book	
		Chapter	Page
To know that there are uncertainties about how global climate change will impact on the UK's future climate	The UK climate today and changes over the past 1000 years	5	71
To understand the impacts of climate change on people and landscapes in UK	The negative effects that climate change is having on the environment and people	6	103
To recognise that there are a range of responses to climate change in the UK at a local and national scale		15	252

⬆ **Figure 15.13** The UK's climate change challenges.

KEY TERMS

Climate change – a long-term movement in the weather patterns and average temperatures experienced by the Earth.

There are uncertainties about how global climate change will impact on the UK's future climate

There are many uncertainties about climate change and its impact on the UK. The belief is that the average temperature that the UK receives is rising and data has been produced to prove the fact. It is also believed that more storms are occurring. There also seems to be periods of dry weather for a number of months then extremely wet weather, which causes flooding, for example the Somerset floods in 2014. The UK coastline has also experienced more storm waves which, coupled with rising sea levels, has caused millions of pounds of damage to coastal defences around the country.

Impacts of climate change on people and landscapes in UK

The impacts of climate change are dealt with on a global scale in the specification, but what will be the impacts for the UK?

● Sea level rises – a rise in sea level caused by climate change would impact on low-lying areas of the UK coastline. Many of the major cities in the country are also ports and therefore their defences will have to be raised and strengthened if the sea level continues to rise. For example, the Thames Flood Barrier will have to be redesigned with higher moving parts.
● Storms and floods – another impact of climate change is that there will be more storms and floods in the UK. This will have a major impact on coastal areas and river valleys, but high winds can also cause tree damage which can be fatal.
● Warmer temperatures – if the temperature of the UK continues to increase there may be changes in the types of crops that are grown. For a number of years farmers in the south of the country have been growing vines and making wine; some are even producing tea.

Responses to climate change in the UK at a local and national scale

By the government

The government has many initiatives to encourage home owners and business to be more energy efficient and to use renewable fuels rather than traditional fossil fuels. For example:

- The renewable heat incentive: home owners who use air or ground source heat pumps to provide heating and water for their homes will receive a grant from the government. This is a payment over seven years to recoup the cost of the heat pump.
- The same incentive will also give home owners a grant if they install thermal hot water heating on the roofs of their homes.
- Businesses and home owners that have solar panels on their roof to produce electricity receive the 'feed in tariff'. The extra electricity they produce feeds into the national grid and the home owner or business is paid for it.
- Electric cars are exempt from road tax; many buses are now electric.
- There are many grants available to enable home owners to insulate their homes for free.

By schools

Many schools are introducing energy-efficient water and central heating systems run from renewable sources such as wind turbines or solar panels. Schools also have notices to switch off lights; some schools have energy prefects who go around the building switching off lights and computers at the end of the school day.

⬆ **Figure 15.14** Wind panels on a house.

By local councils

The UK's target is to cut carbon emissions to fifteen per cent below the 1990 levels by 2010 and twenty per cent by 2020. The government believes that local councils are important in the reduction of carbon emissions as they have an influence on local home owners – fifteen per cent of the UK's carbon emissions are produced by houses. Cutting carbon emissions has been part of local councils' targets since April 2008. To help them meet these targets the government has given them £4 million. The idea is for those local councils that have already introduced ideas to cut carbon emissions to help those that haven't. The most successful councils are Eastleigh Borough Council, the City of London, Barking and Dagenham, Middlesbrough, Woking Borough Council and Worcestershire County Council. These councils have all introduced schemes that have cut carbon emissions. The changes can be as simple as giving away free low-energy light bulbs or as sophisticated as Woking's councils CHP (combined heat and power) scheme.

⬆ **Figure 15.15** Solar panels on a community centre.

By local interest groups

One such group is 'Manchester is my planet'. This group is running a 'pledge campaign' to encourage individuals to reduce their carbon footprint and become involved in a number of green energy projects. The group started in 2005 and works with the local council. There are now over 20,000 individuals who have pledged to work towards a low-carbon future. One of the initiatives is the Green Badge Parking Permit. People who own cars that have been recognised as having low carbon emissions can apply for a Green Badge Parking Permit which allows them to buy an annual parking permit for NCP car parks within Greater Manchester at a 25 per cent discount.

ACTIVITIES

1 Study Figure 15.8 on page 247. Compare the data for immigration and emigration between 2003 and 2013.
2 Find pictures on the internet of greenfield and brownfield developments. Annotate the pictures with the costs and benefits of the different sites.
3 Building on brownfield sites has a number of advantages and disadvantages. Discuss these advantages and disadvantages from the following points of view:
   a) a builder
   b) local residents
   c) a young couple wishing to buy a property.
4 Study Figure 15.9 on page 248. Construct a table which has columns for and against international migration. Put the comments into the correct column.
5 The government gives grants to people who install renewable forms of heat. Research the grants that are available. (Hint: the following website will be useful: www.theccc.org.uk.)

Extension

'National parks are areas of the country that should be preserved and access to them limited.' Discuss this viewpoint with reference to at least two contrasting national parks.

Examination-style questions

1 Study Figure 15.2 which shows the population of the UK from 1971 to 2031 (projected).
   a) State which 10 year period has the fastest rate of growth. (1 mark)
   b) The population of the UK started to grow quickly between 2001 and 2011. Suggest a reason for this increase in the growth of the population. (2 marks)
   c) London had the highest rate of growth of any UK region. Suggest why? (2 marks)
2 Study Figure 15.8 which shows international migration for the UK.
   a) Calculate the median for net migration. (1 mark)
   b) Calculate the range for immigration between 1995 and 2013. (1 mark)
   c) State in which year between 1995 and 2013 net migration was lowest. (1 mark)
   d) Explain why net migration figures might be an unreliable source of information. (4 marks)
3 Use your knowledge and understanding from the rest of your geography course of study to support your answer.
   Discuss the view that protecting coastal areas and settlements on rivers is pointless due to the impact of climate change. (16 marks)

Total: 28 marks

Glossary

Abiotic factors – the physical, non-living environment, such as water, wind, oxygen.

Abrasion – material carried by rivers/waves is thrown against the river bed/cliff face; these particles break off more rocks which, in turn, are thrown against the river bed/cliff face.

Afforestation – the planting of trees in an area that has not been forested before.

Agricultural usage – water used by farmers to water their crops or animals.

Air quality – the relative amounts of air-borne pollutants that can harm human health and the environment, such as sulphur dioxide and nitrogen oxides.

Alluvium – a river deposit of clay, silt and sand.

Altitude – height above sea level.

Ancient woodlands – contain trees that were planted before 1600.

Annual temperature range – the difference between the highest and lowest temperatures of a place in a year.

Anomaly – something that is outside the norm.

Appropriate technology – technology that is suitable for the skill level of the country it is in.

Aquaculture – breeding of fish in pens under controlled conditions.

Aquifers – water-bearing rocks.

Arable farming – growing of cereal crops.

Arête – sharp mountain ridge.

Atmosphere – the gases that surround the Earth.

Backwash – the movement of a wave back down the beach.

Balance of trade – the difference between a country's imports and exports.

Beach nourishment – placing of sand and pebbles on a beach.

Biodiversity – the number of species present in an area.

Biomass – the amount or weight of living or recently living organisms in an area; it can be used as an energy source in biomass boilers.

Biosphere – the part of the Earth and its atmosphere in which living organisms exist or that is capable of supporting life.

Biotic factors – the living organisms found in an area.

Boreal forests – found in areas that have a short growing season; they are made up of tree species that do not lose their leaves when the weather becomes colder.

Bottom-up approaches – these are self-help schemes. The residents of an area are in charge of what happens. They are usually given monetary help and advice on how to improve their houses.

Boulevards – wide, tree-lined streets.

Broad-leaved trees – deciduous trees which lose their leaves in winter, such as oak and elm.

Brownfield site – land within a city that is no longer used; it may contain old factories or housing, or it may have been cleared for redevelopment.

Carbon credits – a permit which allows the holder to emit one ton of carbon dioxide or another greenhouse gas; they can be traded between businesses or countries. Catchment area – area from which a river drains water.

Census – a complete population count for a given area or place taken on a specific date.

Central Business District (CBD) – the centre of the city; it contains the most important shops, businesses and entertainment facilities.

CERF – the United Nations' Central Emergency Response Fund.

Channel shape – the width and depth of the river.

Channelisation – the river channel is made deeper, wider and straighter.

Choropleth map – map in which areas are shaded or patterned in proportion to the measurement of the variable being displayed.

Circulation cells – transfer surplus heat energy from the Equator to the poles.

Climate – the average temperature and precipitation figures for an area.

Climate change – a long-term movement in the weather patterns and average temperatures experienced by the Earth.

Climate graph – graph that shows temperature as a line at the top and rainfall in bars beneath on the same graph.

Coastal flooding – the inundation of land close to the sea by sea water.

Coastal recession – the gradual movement backwards of the coastline, which is the dividing line between the land and the sea.

Colonise – to become established in an area.

Commodity – an item that can be bought or sold.

Composition – what a material is made up of.

Concordant coastline – has rocks that lie parallel to the coastline.

Confluence – the place where two rivers meet.

Coniferous – trees which stay in leaf all year round (evergreens).

Connectivity – the way that a city is connected or linked to other settlements in the UK and to other countries in the world.

Conservation – keeping something as it is; not changing it in any way to preserve it for future generations.

Constructive waves – responsible for deposition in coastal areas and landforms such as beaches, bars and spits.

Core and periphery – the core of a country contains the most wealth while the periphery contains less wealth and, on some occasions, most of the population.

Corrie – an armchair-shaped hollow left after glaciations.

Cortiços – inner-city accommodation for the poor. Families live in one room with shared toilets and cooking facilities. Many of the buildings were previously offices blocks or the homes of the wealthy before they left the inner-city area.

Cottage industry – small-scale production, often in a room of a person's home.

Counter-urbanisation – the movement of people from cities to countryside areas.

Crag – crags form when a glacier meets a hard, resistant rock such as granite or a volcanic plug. The glacier will pass over the hard rock but leave it standing, eroding the softer rock in front of it.

Crystals – a solid material that is arranged in a regular form with definite lines of symmetry.

Cyclone – a low-pressure weather system characterised by inward spiralling winds; also known as a hurricane or typhoon.

Dam – a large, usually concrete, structure built across a river valley to hold back water.

Debris – loose natural material.

Decentralisation – the process of spreading or dispersing power or people away from the central authority.

Deciduous – broad-leaved trees, such as oak and ash, which lose their leaves in the autumn and regrow them each spring.

Deficit – when demand exceeds supply.

Deforestation – the cutting down and removal of all or most of the trees in a forested area.

Degree of urbanisation –the amount of built-up area that has developed in a region.

Deindustrialisation – the reduction of industrial activity in a region.

Demand – a need for something to be sold or supplied.

Density of population – the number of people in an area, usually given as people per square kilometre.

Deposition – process in which sediments, soil and rocks are added to a landform or land mass.

Depression – a low-pressure system that produces clouds, wind and rain.

Desalination – the removal of minerals from salt water to make it drinkable.

Deserts – a barren area of land where little precipitation occurs.

Destructive waves – cause coastal erosion by taking sediment away from coastlines.

Detached house – a house that is not joined to another house.

Detritivore – an animal that feeds on dead plant or animal matter.

Developed country – a country with very high human development (VHHD).

Developing country – a country with low human development (LHD); a poor country.

Development – an improvement in the quality of life for the population of a country.

Development gap – the difference between the parts of the world that have wealth and the parts of the world that do not.

Direct and indirect tourism jobs – direct tourism jobs are jobs such as a waitress in a hotel; indirect jobs would include farmers providing food for the hotel.

Discharge – the amount of water passing a specific point at a given time, measured in cubic metres per second.

Discordant coastline – has bands of rocks that lie at right angles to the coast.

Disposable income – the money people have left to spend after essential goods and services have been paid for.

Distribution – where something is located.

Diurnal – daily changes.

Domestic water usage – use of water by households.

Donor countries – countries that give aid.

Drift – all sediments deposited by glaciers.

Drought – a period of below-average precipitation resulting in prolonged shortages in waters supply.

Drumlin – elongated glaciated landform formed of till.

Dynamic landform – a landform that is changing.

Economic water scarcity – that applies to areas that lack the capital or human power to invest in water sources and meet local demand; water is often available for people who can pay for it but not for the poor.

Ecosystem – a community of plants and animals and their non-living environment.

Ecotourism – travel to natural areas that does no damage, conserving the environment and improving the well-being of local people.

Edge- and out-of-town shopping – shops or facilities located away from the city centre, on the edge of cities.

Emerging country – a country with high and medium development (HMHD).

Emigration – the process of moving out of a country.

Employment structure – the numbers of people employed in each sector of industry.

Enclosure Acts – a series of Acts of Parliament between 1750 and 1860 which stopped villages using the open fields and commons that they had been allowed to use for centuries. This meant that villagers could not make a living and had to move to find a better life. Many went to live in industrial towns.

Energy consumption – how much energy is used.

Energy mix – the way that countries use energy in different proportions.

Energy production – how much energy is being made.

Energy-efficient housing – less energy is used to provide the same level of heat or power; homes are well insulated.

Englacial debris – rock material that is carried in the main body of the glacier.

Enhanced greenhouse effect – also called climate change or global warming. It is the impact on the global climate of the increased amounts of carbon dioxide and other greenhouse gases that humans have released into the Earth's atmosphere since the Industrial Revolution.

Environmental sustainability – the ability to maintain the qualities that are valued in the physical environment

Erosion – the wearing away of materials by a moving force, such as a river, teh sea or ice.

Estuary –transition zone where rivers meet the sea.

Ethnicity – relates to a group of people who have a common national or cultural tradition.

Eutrophication – the growth of algae on water courses due to the increase in chemical fertilisers being used on the land.

Evapotranspiration – the movement of water from the Earth, by evaporation and plant transpiration, to the atmosphere.

Exploitation – the act of using natural resources.

Extraction – any activity that withdraws resources from nature, for example mining.

Favela – homes for the poor which can be found anywhere in the city. They are made from waste materials and have no water supply, electricity or toilets.

FEMA – the USA's Federal Emergency Management Agency.

Fetch – the distance over which the wind blows over open water.

Finite resource – a resource that will eventually run out.

Floodplain – the flat area of land that the river spills on to when it bursts it banks.

Floodplain zoning – land that is close to the river is seen as low value because of flood risk; it is used for recreation, for example sports fields. Housing areas would be further away on more valuable land that is less likely to flood.

Fluvio-glacial material – rocks and other debris deposited by melting ice or glacial streams.

Food chain – a series of steps by which energy is obtained and used by living organisms.

Food web – a network of food chains by which energy and nutrients are passed from one species to another; it is essentially 'who eats what'.

Forestry – the practice of planting, managing and caring for large areas of trees.

Fossil fuel – a naturally occurring fuel such as coal, oil and natural gas (methane) formed from the remains of dead organisms over millions of years.

Fossil – the remnants of prehistoric organisms, such as a fish skeleton or a leaf imprint, which have become embedded in a rock.

Fracking – the process of drilling down into gas-bearing rock before blasting it with water, sand and chemicals at high pressure to release the gas inside the rock layers; the gas then flows out through the top of the well.

Free market economy – free competition between producers; prices are determined by supply and demand.

Freeze–thaw weathering (frost action) – occurs when water gets into cracks in rocks; it expands when it turns into ice, which puts pressure on the rock around it.

Fresh water – water that contains less than 1000 milligrams per litre of dissolved solids, most often salt.

Function –the function of a tropical rainforest is its ecosystem and how it works.

Geology – the science that deals with the physical structure of the Earth, its history and how it changes.

Geopolitical – the influence of factors such as geography and economics on the politics and foreign policy of a state.

Gersmehl model – model of the mineral nutrient cycle

Glacial trough – U-shaped valley formed by a valley glacier.

Global atmospheric circulation – the worldwide movement of air which transports heat from tropical to polar latitudes.

Global climate change – see climate change.

Globalisation – the way that companies, ideas and lifestyles are spread around the world.

GOAD maps – detailed street maps including individual buildings and their uses.

Gradient – the slope over which the river loses height.

Granite batholith – a large area of igneous rock which forms below the Earth's surface.

Green belt land – an area around a city composed of farmland and recreational land. There are strict controls on the development of this land; its purpose is to control the growth of cities.

Greenfield development – when houses and other buildings are built on land at the edge of a city that has never been built on before.

Greenhouse effect – see enhanced greenhouse effect

Greenhouse gases – gases that trap heat within the Earth's atmosphere.

Gross domestic product (GDP) – this is the value of all the goods and services produced in a country during a year, in US$.

Gross national income (GNI) – the value of all the goods and services produced in a country and from its exports during a year, in US$.

Ground moraine – disorganised piles of rocks of various shapes, sizes and of differing rock types laid down by a glacier.

Groyne – a rigid structure that interrupts water flow and limits the movement of sediment.

Hard engineering – this method of coastal and river management that involves major construction work, for example sea walls and dams.

Heathland – tends to be open countryside in lowland areas. The plants are small shrubs, such as heather and gorse, with a few silver birch trees.

Hemisphere – half of the Earth. The northern hemisphere is above the Equator, the southern hemisphere is below the Equator.

Hibernate – to spend the winter in close quarters in a dormant (sleeping) condition.

Highland boundary fault – a fault line across Scotland from west to east that divides two very different geological areas. It separates the highlands of Scotland from the lowlands of Scotland.

Human causes – any occurrence that is created by humans.

Human Development Index (HDI) – a measurement of life expectancy, access to education and gross national income per capita used to assess how much progress a country has made (see http://hdr.undp.org).

Humanitarian aid – help given after a natural disaster to save lives and reduce suffering.

Hydroelectric power (HEP) – energy produced by water turning a turbine to produce electricity.

Hydrograph – a graph showing rainfall and river discharge over a specific period of time.

Ice core – a core sample removed from an ice sheet.

Ice fairs – amusements held on the River Thames during the Little Ice Age.

Igneous rock – formed from molten rock called magma that is found inside the Earth.

Imbalance – something that is not equal.

Immigration – the process of moving into a country.

Index of political corruption – perceived levels of public sector corruption.

Industrial usage – use of water by factories or the companies that produce energy; it also applies to water used in offices and schools.

Informal sector – people who set up informal businesses such as selling products on the street; they do not pay taxes or rent proper business premises.

Infrastructure – in this instance it refers to pipes that bring water to our homes.

Inner city – the central area of a major city.

Inorganic material – something that was never living matter.

Interception – when trees stop precipitation hitting the ground surface.

International aid – assistance, usually financial, from another country.

International migration – the movement of people from one country to another with the intention of staying there for at least a year.

International Monetary Fund (IMF) – financial institution set up in 1945 to promote international trade. It pools funds from 188 member nations who can then make withdrawals when their economy is in difficulty.

Internet shopping – people buying goods online, which allows them to shop from home.

Irrigation – the artificial watering of land for farming.

Inter Tropical Convergence Zone (ITCZ) – the area encircling the Earth near the Equator where the northeast and southeast trade winds come together.

Karst limestone – a landscape formed from the dissolution of soluble rocks, such as limestone.

Latitude – the distance north or south of the Equator. It is measured in degrees with the maximum being 90°N or 90°S.

Lee slope – the side of a landform facing away from the direction of the ice flow.

Levee – depositional landforms that form when the river is in flood or during times of heavy rainfall.

Limiting factors – factors that limit biodiversity/population size, such as temperature, moisture, light and nutrients; these factors are in abundance in tropical rainforests, which accounts for their high biodiversity.

Literacy – the ability to read and write.

Litter – decomposing leaf and other organic debris found on forest floors.

Long profile – a slice through the river from source to mouth that shows the changes in height of the river's course.

Magma – molten rock under the surface of the Earth.

Major city – a city with a population of at least 400,000.

Major urban centre – an area that has a high population density and is made up of houses, industrial buildings, factories and transport routes. These areas are referred to as built-up areas.

Managed retreat – when the land is allowed to flood in a controlled way.

Maritime – influenced by the sea.

Mass movement – when material moves down a slope due to the pull of gravity.

Measures of inequality – these are ways of measuring how equal people are within a country or between countries.

Median age range – the middle age range, if all ages are put in a line.

Metamorphic rock – rock formed when igneous or sedimentary rocks are put under great pressure or are close to a source of heat.

Metered households – meters measure the amount of water that a household uses and the home owner pays for the exact amount of water consumed.

Methane – fossil methane, which provides approximately 30 per cent of methane released into the atmosphere, was formed underground many years ago. It comes to the surface when fossil fuels are mined. Methane is a greenhouse gas; this means that it can trap heat within the Earth's atmosphere. It makes up twenty per cent of the greenhouse gases in the atmosphere and is twenty times more potent than carbon dioxide.

Migration – the movement of people from one area to another with the intention of staying there for at least a year.

Milankovitch cycles – long-term variations in the orbit of the Earth which result in changes in climate.

Mineral – a solid, naturally occurring non-living substance, such as coal or diamonds.

Mixed farming – a farm which has cereal crops and animals.

Monoculture – the growing of one crop on large areas of land.

Moorland – land which is not intensively farmed. It is found in upland areas of the UK and tends to have acidic, peaty soils. The plants are small shrubs such as heather; there are few trees.

Mouth – where the river ends, either when it joins another river or meets the sea.

National migration – the permanent movement (for at least one year) of people from one area of a country to another.

National park – an area of countryside that is protected because of its natural beauty and managed for visitor recreation.

Nationalised – to convert to a business or industry from private to government ownership.

Natural increase – when population numbers show a positive difference between the birth rate and the death rate.

Natural resources – materials that are provided by the Earth that people make into something that they can use.

Net migration – the difference between the number of people who are coming into a country and the number of people leaving a country.

NGO – non-governmental organisation; a not-for-profit organisation that is not under government control. They are usually set up by private individuals and can be funded by donations or governments.

Non-renewable energy – energy sources that, once they have been used, can never be used again.

Nomadic herdsmen – people raising animals for their own food; they move around and have no fixed land.

North–South divide – a virtual socioeconomic and political line on the globe which splits the developed and wealthy countries in the 'North' from the poorer developing countries in the 'South'.

Nutrient cycle – the movement and exchange of organic and inorganic material into living matter.

Ocean current – a continuous, directed movement of ocean water. The currents are made from forces acting on the water such as the wind, different temperatures and the Earth's rotation.

Offshore oilfield – oil drilled from under the sea.

Offshore reefs – concrete blocks or natural boulders sunk offshore to alter wave direction and dissipate wave energy.

Onshore oilfield – oil drilled on land.

Onshore wind farm – an area of land that is covered by wind turbines.

Organic material – something that was once living.

Overfishing – when fish stocks are reduced to below acceptable levels.

Overgrazing – grass that is grazed so heavily that the vegetation is damaged and the ground becomes liable to erosion.

Overpopulation – too many people living in an area for the area to support.

Owner-occupied – the house is owned by the people who live in it or they are buying it with a mortgage.

Pastoral farming – the rearing of sheep, cattle, pigs or any other animals on a farm.

Paulistanos – the name given to the residents of Sao Paulo.

Per capita – average per person.

Physical causes – any occurrence that is natural.

Physical water scarcity – term that applies to dry, arid regions where fresh water naturally occurs in low quantities.

Plucking – a form of glacial erosion.

Political Corruption Index – perceived levels of public sector corruption.

Pollen records – examining accumulated pollen to determine how the plant species in an area has changed.

Population density – the number of people living in a certain area of land.

Population distribution – the pattern showing how population is spread over an area.

Poverty line – the minimum level of income deemed adequate in a particular country.

Precipitation – any form of moisture that reaches the Earth; rain, snow, etc.

Prevailing wind – the direction from which the wind usually blows. In the UK it is the southwest.

Primary data source – first-hand evidence collected by the researcher themselves.

Primary sector – made up of extractive industries such as farming, fishing, forestry and mining. Countries which are developing have high numbers of people employed in this sector.

Private investment – money put into businesses by private financial backers.

Public buildings – buildings owned by the council that serve the residents of the city, such as a library.

Public investment – money put into businesses by the government.

Pull factors – the reasons why people are attracted to a city.

Push factors – the reasons why people want to leave rural areas.

Pyramidal peak – the sharpened top of a mountain formed from the back walls of three or more corries.

Qualitative techniques – techniques in which information is gained through observation; it usually involves a description of a feature.

Quality of life – the general well-being of individuals and societies.

Quantitative techniques – data collection techniques that record statistical data and/or measurements and are carried out in the field.

Quaternary sector – financial services and telecommunications.

Random sampling – this is where things are chosen by chance; a random number table could be used.

Rate of urbanisation – the speed at which settlements are built.

Recycling – the process of converting waste materials into new products.

Refugee – a person who has been forced to leave their country in order to escape war, persecution or natural disaster.

Regeneration – renewing or revitalising an existing area.

Region – an area within a country.

Renewable energy – energy that comes from sources that can be reused or replenished and therefore will not run out.

Reservoir – large area of water created after the flow of a river has been controlled, often by building a dam.

Residential – an area used for housing.

Resistant – strong rocks that can withstand weathering and erosion.

Resource – a stock or supply of something that is useful to people.

Re-urbanisation – the movement of people back into urban areas, usually after a city has been modernised.

Rip rap – the use of rocks or other material used to defend shorelines against erosion.

Roche moutonnée – glacial feature formed from an outcrop of more resistant rock that was in the path of the ice as it moved across the upland area.

Rural depopulation – the movement of people from rural to urban areas.

Rural to urban migration – the movement of people from the countryside to towns and cities.

Saltation – small pieces of shingle or large grains of sand are bounced along a river or sea bed.

Sao Paulo city area – the inner built-up area of Sao Paulo, which has approximately 11 million inhabitants.

Sao Paulo metropolitan area – the whole of the built-up area; it includes Sao Paulo and a number of nearby cities; it has approximately 19 million inhabitants.

Savannah ecosystem – an area of grassland which has a few shrubs and trees; it can be found in tropical areas.

Seasonal – relating to a particular season of the year.

Seasonal imbalance – when rainfall and demand for water are unevenly distributed throughout the year.

Secondary data source – evidence collected by someone else.

Secondary sector – manufacturing industries; the number of people employed in this sector increases as a country develops.

Sediment – naturally occurring material that is broken down by processes of weathering and erosion.

Sedimentary rock – rock formed in layers from weathered or eroded rock debris that has been transported and deposited.

Semi-detached house – house that is joined on just one side to another house.

Shale rock – a fine-grained sedimentary rock.

Short rotation coppice – trees, usually willow, grown specifically to be used fuel for biomass boilers for domestic heating or power stations. They are planted densely and harvested on two- and five-year cycles.

Silt – a granular material of a size somewhere between sand and clay.

Site – the land that the settlement is built upon.

Situation – where the settlement is compared to physical and human features around it.

Slave triangle – describes a three-part journey: ships left British ports such as Bristol with goods such as cloth and guns, they sailed to Africa where these goods were sold and enslaved people were bought. The enslaved people were taken to the Caribbean and sold. Goods such as sugar were then bought with the money and brought back to England.

Social priority housing – houses that are owned by a housing association and rented to people who cannot afford to buy their own home.

Socialism – a political model based on the belief that a country's people should own its means of production and regulate its political power.

Soft engineering – this method of coastal and river management works or attempts to work with the natural processes occurring on the coastline, for example beach nourishment and river washlands. They tend to be visually unobtrusive and do not involve major construction work.

Soil – the top layer of the earth in which plants grow; it contains organic and inorganic material.

Solar power – renewable energy from the Sun.

Solar variation – variations in the Sun's activity.

Solution – some minerals are dissolved in sea or river water; it can change the colour of the water due to the minerals that are present.

Source – the start of a river.

Source region – a large area of the Earth's surface where the air has a uniform temperature and humidity.

Spatial variations – differences in something on the Earth's surface, for example, differences in wealth in a country.

Spur – a long narrow tongue of land that drops from high ground to lower ground.

Squatter settlements – a shanty town made from waste materials.

SSSI – site of special scientific interest that is protected because of its unique habitat.

Stack – a piece of rock that stands in the sea; it used to be joined to the headland.

Stakeholder – someone with an interest in what occurs.

Stoss end – the side of a landform facing the direction of the ice flow.

Stratified sampling – the population is divided into sections known as 'strata'; the sampler ensures that the same amount of data is taken from each stratum. For example, the strata could be roads in the CBD: the sampler would ensure that they interviewed ten people on each road.

Structure – the structure of a tropical rainforest is the layers of plants and animals in the forest.

Suburbanisation – the growth of a town or city into the surrounding countryside, which usually joins it to villages on its outskirts making one large built-up area.

Suburbs – a residential area or a mixed use area within commuting distance of a city.

Supply – the amount of something that can be provided or sold.

Supraglacial debris – rock material that is carried on the surface of a glacier.

Surplus – when supply exceeds demand.

Suspension – small particles such as sand and clays are carried in the water; this can make the water look cloudy, especially during storms or after heavy rainfall when the sea/river has lots of energy.

Sustainable management – using energy resources in a way which ensures that they are not exploited and will hopefully be able to meet the needs of future generations.

Swash – the forward movement of a wave.

Systematic sampling – this means the sample is chosen according to an agreed interval, for example, every fifth person who walks passed is interviewed.

Tarn – a small mountain lake in a corrie.

Tectonic activity – the movement of the Earth's plates.

Temperate forests – forests found in temperate regions; the main characteristics include: wide leaves, large and tall trees and non-seasonal vegetation.

Terminal moraine – deposit of rocks found at the furthest (end) point reached by a glacier.

Terraced house – houses that are joined on each side to the house next to them; their front doors usually open straight on to the street.

Tertiary sector – service industries and jobs such as teaching; very few people are employed in this sector in a developing country.

Texture – the feel and appearance of a material.

Throughflow – when water travels through soil towards a river.

Till – all the material deposited directly by a glacier, which is collectively known as moraine.

Top-down approaches – this is when the government improves an area and expects people to move into the housing they have provided. Sometimes the government borrows large sums of money from other countries to pay for the scheme.

Total annual rainfall – the sum of all the rainfall that falls in a year in an area.

Traction – large sediment such as pebbles roll along the river/sea bed.

Trade winds – a wind that blows steadily from the tropics towards the Equator. In the northern hemisphere it is from the northeast and in the southern hemisphere from the southeast.

Transnational corporation (TNC) – large company that has its headquarters in one country and branches all over the world.

Transpiration – evaporation of moisture from the leaves of a plant.

Tree rings – annual growth rings seen in a horizontal cross-section.

Tributary – a minor river joining on to the main river.

Tropical forests – forests within the tropics which experience high temperatures and significant rainfall.

Troposphere – the lowest layer of the atmosphere. It is thicker at the Equator (approximately 20 km) than at the poles (approximately 10 km).

Tundra – an area where the tree growth is hindered by low temperatures and short growing seasons.

Two-speed economy – the South East has a faster rate of growth than the rest of the UK.

Unmetered households – the water company estimates how much water the householder will use and charges for this amount.

Urbanisation – the increase in the number of people living in towns and cities compared to the number of people living in the countryside.

Urban-rural fringe – the area on the outskirts of the city.

Valley profile – a slice across a river showing the changes in height across the valley.

Velocity – the speed of the river.

Volcanic plug – a volcanic cone from an extinct volcano.

Volcanism – volcanic activity or phenomena.

Volume – the amount of water in the river.

Wace-cut platform –narrow flat area found at the base of a cliff that was created by the erosion of waves.

Washlands – the river is allowed to flood these areas; it could be farmland or recreational land close to settlements.

Water cycle – the closed system in which water moves between the atmosphere, the oceans and land.

Water deficit – water demand exceeds water supply.

Water quality – a measure of the levels of pollutants in water that can harm human health and the environment.

Water scarcity – the lack of sufficient available water resources to meet the demands within a region.

Water supply – the provision of water by public utilities, commercial organisations, communities or individuals.

Water surplus – when water supply exceeds water demand.

Water usage – the amount of water used.

Weather – the day-to-day changes in temperature and precipitation.

Weathering – the breaking down or dissolving of rocks and minerals on Earth's surface.

Wetlands – areas of low-lying land that is predominantly wet and boggy. Some wetland areas have been drained, such as the Somerset Levels and the Fens. The term 'wetland' also refers to small ponds and river estuaries.

Wind power – renewable energy from the wind.

Woodland – tree-covered area.

World Bank – an international financial institution that provides loans to developing countries.

Index